Reviews of Environmental Contamination and Toxicology

VOLUME 190

Reviews of Environmental Contamination and Toxicology

Continuation of Residue Reviews

Editor
George W. Ware

Associate Editor
David M. Whitacre

VOLUME 190

Springer

Springer
New York: 233 Spring Street, New York, NY 10013, USA
Heidelberg: Postfach 10 52 80, 69042 Heidelberg, Germany

ISBN 978-1-4419-2274-8 e-ISBN 978-0-387-36903-7
ISSN 0179-5953

Printed on acid-free paper.

springer.com

Foreword

International concern in scientific, industrial, and governmental communities over traces of xenobiotics in foods and in both abiotic and biotic environments has justified the present triumvirate of specialized publications in this field: comprehensive reviews, rapidly published research papers and progress reports, and archival documentations. These three international publications are integrated and scheduled to provide the coherency essential for nonduplicative and current progress in a field as dynamic and complex as environmental contamination and toxicology. This series is reserved exclusively for the diversified literature on "toxic" chemicals in our food, our feeds, our homes, recreational and working surroundings, our domestic animals, our wildlife and ourselves. Tremendous efforts worldwide have been mobilized to evaluate the nature, presence, magnitude, fate, and toxicology of the chemicals loosed upon the earth. Among the sequelae of this broad new emphasis is an undeniable need for an articulated set of authoritative publications, where one can find the latest important world literature produced by these emerging areas of science together with documentation of pertinent ancillary legislation.

Research directors and legislative or administrative advisers do not have the time to scan the escalating number of technical publications that may contain articles important to current responsibility. Rather, these individuals need the background provided by detailed reviews and the assurance that the latest information is made available to them, all with minimal literature searching. Similarly, the scientist assigned or attracted to a new problem is required to glean all literature pertinent to the task, to publish new developments or important new experimental details quickly, to inform others of findings that might alter their own efforts, and eventually to publish all his/her supporting data and conclusions for archival purposes.

In the fields of environmental contamination and toxicology, the sum of these concerns and responsibilities is decisively addressed by the uniform, encompassing, and timely publication format of the Springer triumvirate:

Reviews of Environmental Contamination and Toxicology [Vol. 1 through 97 (1962–1986) as Residue Reviews] for detailed review articles concerned with any aspects of chemical contaminants, including pesticides, in the total environment with toxicological considerations and consequences.

Bulletin of Environmental Contamination and Toxicology (Vol. 1 in 1966) for rapid publication of short reports of significant advances and discoveries in the fields of air, soil, water, and food contamination and pollution as well as methodology and other disciplines concerned with the introduction, presence, and effects of toxicants in the total environment.

Archives of Environmental Contamination and Toxicology (Vol. 1 in 1973) for important complete articles emphasizing and describing original experimental or theoretical research work pertaining to the scientific aspects of chemical contaminants in the environment.

Manuscripts for *Reviews* and the *Archives* are in identical formats and are peer reviewed by scientists in the field for adequacy and value; manuscripts for the *Bulletin* are also reviewed, but are published by photo-offset from camera-ready copy to provide the latest results with minimum delay. The individual editors of these three publications comprise the joint Coordinating Board of Editors with referral within the Board of manuscripts submitted to one publication but deemed by major emphasis or length more suitable for one of the others.

Coordinating Board of Editors

Preface

The role of *Reviews* is to publish detailed scientific review articles on all aspects of environmental contamination and associated toxicological consequences. Such articles facilitate the often-complex task of accessing and interpreting cogent scientific data within the confines of one or more closely related research fields.

In the nearly 50 years since *Reviews of Environmental Contamination and Toxicology* (formerly *Residue Reviews*) was first published, the number, scope and complexity of environmental pollution incidents have grown unabated. During this entire period, the emphasis has been on publishing articles that address the presence and toxicity of environmental contaminants. New research is published each year on a myriad of environmental pollution issues facing peoples worldwide. This fact, and the routine discovery and reporting of new environmental contamination cases, creates an increasingly important function for *Reviews.*

The staggering volume of scientific literature demands remedy by which data can be synthesized and made available to readers in an abridged form. *Reviews* addresses this need and provides detailed reviews worldwide to key scientists and science or policy administrators, whether employed by government, universities or the private sector.

There is a panoply of environmental issues and concerns on which many scientists have focused their research in past years. The scope of this list is quite broad, encompassing environmental events globally that affect marine and terrestrial ecosystems; biotic and abiotic environments; impacts on plants, humans and wildlife; and pollutants, both chemical and radioactive; as well as the ravages of environmental disease in virtually all environmental media (soil, water, air). New or enhanced safety and environmental concerns have emerged in the last decade to be added to incidents covered by the media, studied by scientists, and addressed by governmental and private institutions. Among these are events so striking that they are creating a paradigm shift. Two in particular are at the center of ever-increasing media as well as scientific attention: bioterrorism and global warming. Unfortunately, these very worrisome issues are now super-imposed on the already extensive list of ongoing environmental challenges.

The ultimate role of publishing scientific research is to enhance understanding of the environment in ways that allow the public to be better informed, The term "informed public" as used by Thomas Jefferson in the age of enlightenment conveyed the thought of soundness and good judgment. In the modern sense, being "well informed" has the narrower meaning of having access to sufficient information. Because the public still gets most of its information on science and technology from TV news and reports, the role for scientists as interpreters and brokers of scientific information to the public will grow rather than diminish.

Environmentalism is the newest global political force, resulting in the emergence of multi-national consortia to control pollution and the evolution of the environmental ethic. Will the new politics of the 21st century involve a consortium of technologists and environmentalists, or a progressive confrontation? These matters are of genuine concern to governmental agencies and legislative bodies around the world.

For those who make the decisions about how our planet is managed, there is an ongoing need for continual surveillance and intelligent controls, to avoid endangering the environment, public health, and wildlife. Ensuring safety-in-use of the many chemicals involved in our highly industrialized culture is a dynamic challenge, for the old, established materials are continually being displaced by newly developed molecules more acceptable to federal and state regulatory agencies, public health officials, and environmentalists.

Reviews publishes synoptic articles designed to treat the presence, fate, and, if possible, the safety of xenobiotics in any segment of the environment. These reviews can either be general or specific, but properly lie in the domains of analytical chemistry and its methodology, biochemistry, human and animal medicine, legislation, pharmacology, physiology, toxicology and regulation. Certain affairs in food technology concerned specifically with pesticide and other food-additive problems may also be appropriate.

Because manuscripts are published in the order in which they are received in final form, it may seem that some important aspects have been neglected at times. However, these apparent omissions are recognized, and pertinent manuscripts are likely in preparation or planned. The field is so very large and the interests in it are so varied that the Editor and the Editorial Board earnestly solicit authors and suggestions of underrepresented topics to make this international book series yet more useful and worthwhile.

Justification for the preparation of any review for this book series is that it deals with some aspect of the many real problems arising from the presence of foreign chemicals in our surroundings. Thus, manuscripts may encompass case studies from any country. Food additives, including pesticides, or their metabolites that may persist into human food and animal feeds are within this scope. Additionally, chemical contamination in any manner of air, water, soil, or plant or animal life is within these objectives and their purview.

Manuscripts are often contributed by invitation. However, nominations for new topics or topics in areas that are rapidly advancing are welcome. Preliminary communication with the Editor is recommended before volunteered review manuscripts are submitted.

Tucson, Arizona G.W.W.

Table of Contents

Rev Environ Contam Toxicol 190:1–41

Dioxin Formation from Waste Incineration

Takayuki Shibamoto, Akio Yasuhara, and Takeo Katami

Contents

I. Introduction

Adverse effects of dioxins—polychlorinated dibenzo-*p*-dioxins (PCDDs), polychlorinated dibenzofurans (PCDFs), and coplanar polychlorinated biphenyls (PCBs)—on human health have been known for many years. In particular, PCDDs and PCDFs have received much attention recently, not only from environmental scientists but also by the public because they are

Communicated by Ronald S. Tjeerdema.

T. Shibamoto (✉)
Department of Environmental Toxicology, University of California, Davis, CA 95616, U.S.A.

A. Yasuhara
Environmental Preservation Center, Tokyo University of Science, 12-1 Funagawara-machi, Ichigaya, Shinjuku, Tokyo 162-0826, Japan

T. Katami
Gifu Prefectural Institute of Health and Environmental Science, 1-1 Naka-Fudogaoka, Kagamigahara, Gifu, 504-0838, Japan

formed during the incineration of industrial wastes (Coulston and Pocchiari 1983). Consequently, there is a pressing need to find the formation mechanisms or reaction pathways of these chlorinated chemicals to reduce their environmental contamination. However, the formation mechanisms of these chlorinated chemicals are not yet completely understood because their reaction mechanisms are extremely complex (Huang and Buekens 1995; Fängmark et al. 1994).

One of the major factors influencing dioxin formation from an incinerator is the presence of chlorides in combustion materials. There are numerous reports on formation of dioxins from chloride-containing materials, such as polyvinyl chloride, on combustion (Giugliano et al. 1989; Carroll 2001; Yasuhara et al. 2001; Katami et al. 2002). Also, dioxin formation has been reported from flammable materials mixed with inorganic chlorides, such as NaCl, on incineration (Yasuhara et al. 2001, 2002, 2003; Luthe et al. 1998; Katami et al. 2000). According to a study using a fluidized-bed furnace, there was a significant correlation between the chloride content of combustion materials containing inorganic chlorides or organic chlorides and dioxin formation (Hatanaka et al. 2000). However, trace amounts of dioxin formation were observed from the combustion of nonchloride plastics (Katami et al. 2002). Formation of dioxins from combustion of flammable materials is inevitable because trace amounts of NaCl particles are present in the atmosphere. For example, dioxin formation was observed from a nonchloride plastic, polystyrene, on combustion (Yasuhara et al. 2002).

Combustion temperature plays an important role in dioxins formation in an incinerator (Addink and Olie 1995; Stanmore 2004). Appropriate management of combustion conditions, including temperature, is an important factor to reduce dioxin formation during incineration of domestic garbage. According to an experiment done using an experimental-scale furnace, dioxins were formed under 800°C from materials containing inorganic or organic chlorides (Kawabata et al. 2002). However, dioxins were reportedly degraded by 99.9999% at 1,000°C when burned with 1-sec residence time (Nagata 1992). Consequently, the combustion temperatures of an incinerator for domestic garbage have been set above 850°C to prevent the formation of dioxins (PCDDs, PCDFs, and coplanar PCBs) in Germany, France, England, and Japan (Fukunaga 1998). In this review, therefore, the amounts of dioxins formed from various materials with either inorganic chloride or organic chloride combusted under different temperatures in an incinerator are discussed.

There have been many reports on investigations of dioxin formation from waste materials combusted in a small-scale electric furnace equipped with quartz tubing. However, this type of incinerator is not consistent with actual furnaces used to incinerate domestic wastes, and, consequently, the results obtained using this small-scale electric furnace are not always comparable to those obtained from an actual incinerator. It is difficult to use a large-scale incinerator for experiments because even preparing combustion samples and changing the firebricks of the inside walls for each combustion are almost impossible. The small-scale incinerator used for the experiments in the core references of this review did not require tedious sample preparation for combustion experiments. Also, the firebricks were easily changed on the inside walls of this incinerator for each experiment. Therefore, the data obtained from this incinerator are comparable to those obtained from an actual incinerator.

II. Sample Incineration to Investigate Dioxin Formation

There are many reports on dioxin formation from various materials, including newspaper (Katami et al. 2000; Yasuhara et al. 2001), plastics (Katami et al. 2002; Yasuhara et al. 2002, 2005a,b, 2006), wood (Yasuhara et al. 2003; Katami et al. 2004a), and domestic garbage (Katami et al. 2004b), incinerated a under-controlled conditions.

A. Incinerator Description

A typical incinerator designed to combust various samples reported in the core articles of this review is shown in Fig. 1 (Katami et al. 2002). The combustion chamber was made of firebrick with a volume of $0.22\,m^3$. The area of the grate was $0.19\,m^2$. The inlet for combustion samples was 0.35 m

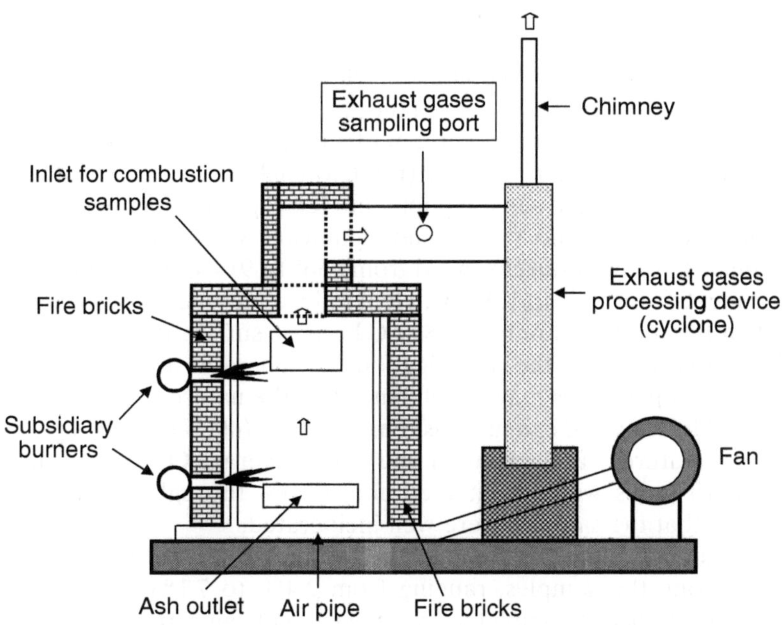

Fig. 1. Typical incinerator used to combust various samples reported in core articles of this review.

(height) × 0.40 m (width). The incinerator was equipped with two subsidiary combustion burners (Kato Burner, Gifu, Japan), with a heat supply of 30,000 kcal/hr to completely combust samples. Firebricks of the inside walls were changed for each experiment to avoid contamination from previous experiments. Combustion gases were exhausted through a chimney after dust was removed by a cyclone (960 m^3/hr) equipped at the outlet of the combustion chamber. Combustion temperatures were measured at the center of the chamber and at the grate. Subsidiary burners were turned on 2 hr before combustion of samples to maintain a constant temperature. Air was supplied through inlets located at the four corners of the chamber. Air flow rate was adjusted with a flow rate control valve.

B. Temperature, O_2, CO, CO_2, and HCl Measurement in Exhaust

The combustion chamber and flame temperatures were measured by a LK-1,200 thermocouple conductor interfaced to a CT-1310 digital thermometer (Custom, Tokyo, Japan). Pretreatment for water removal from exhaust gas was conducted by a PS-200SCR Scrubber (Horiba, Kyoto, Japan). Continuous measurement of CO, CO_2, and O_2 in exhaust gas was performed by a Horiba PG-230 Gas Analyzer (Horiba, Kyoto, Japan). HCL concentration was measured by a Yokogawa IC-7000S ion chromatograph (Yokogawa Analytical Systems, Tokyo, Japan) after exhaust gases were purged into water. The chloride content in the combustion samples was measured by a TOX-100 total organic halogen analyzer (Dia Instruments, Chigasaki, Japan).

Tables 1–7 show typical combustion chamber conditions and concentrations of O_2, CO, CO_2, and Cl ions in the exhaust gases from various samples incinerated, along with the Cl content of each sample. These tables present the results reported in 10 articles (refer to each table), in which samples were incinerated in the same equipment (see Fig. 1). Chamber conditions were kept constant and showed no significant variations among the 10 experiments. Temperatures ranged from 456° to 919°C and grate temperatures ranged from 539° to 1,076°C. Generally, experiments were conducted at a chamber temperature below 800°C because sufficient dioxin formation could be expected for the investigation (Yasuhara et al. 2003). Chamber and grate temperatures may be influenced by the nature of the combusting samples. The range of exhaust gas temperatures (408°–667°C) was smaller than temperatures of the chamber and grate. Average O_2 concentrations in the exhaust gases were maintained above 10% through all experiments, indicating that the samples were well incinerated.

Average CO_2 concentrations in the exhaust gases were generally consistent among the samples, ranging from 2.4% to 7.1%, also suggesting that incineration was complete. Average CO concentrations varied significantly. The overall profile of CO in the exhaust gases is the higher the temperature the lower the concentration. It is difficult to explain CO formation

Table 1. Combustion Chamber Conditions and Concentrations of O_2, CO, CO_2, and Cl Ion in Exhaust Gases from Incinerated NaCl, KCl, or $CaCl_2$ Impregnated Newspapers.

Combustion sample	Chlorine cont (wt%)	Avg chamber temp (°C)	Avg grate temp (°C)	Avg exhaust gas temp (°C)	Avg amount dry exhaust gas (m^3/hr)	Avg O_2 conc (%)	Avg CO_2 conc (%)	Avg CO conc (ppm)	Cl ion conc (mg/m^3 N)	References
Newspaper (NP) alone	0.0064	653	—	526	918	17.2	2.4	1,000	ND	Yasuhara et al. 2001
NaCl-impregnated (I) NP A	3.1	510	—	415	995	16.7	3.4	1,022	23	Yasuhara et al. 2001
NaCl-I NP B	0.99	578	594	461	225	15.0	4.5	251	44	Yasuhara et al. 2002
NaCl-I NP C	1.88	747	698	596	241	7.7	9.9	140	55	Yasuhara et al. 2002
NaCl-I NP D	4.08	751	721	612	245	9.2	8.9	25	88	Yasuhara et al. 2002
NaCl-I NP E	4.08	791	917	623	230	12.5	6.0	>2.0	83	Yasuhara et al. 2002
KCl-I NP	1.25	689	595	455	225	15.3	4.3	271	79	Yasuhara et al. 2002
$CaCl_2$-I NP	0.79	653	664	475	232	15.3	4.4	129	65	Yasuhara et al. 2002

Conc, concentration; avg, average; temp, temperature; —, not reported; ND, not detected.

Table 2. Combustion Chamber Conditions and Concentrations Of O_2, CO, CO_2, and Cl Ion in the Exhaust Gases from Incinerated Newspapers Impregnated with NaCl and Various Inorganic Metal Chlorides.

Combustion sample	Chlorine cont (wt%)	Avg chamber temp (°C)	Avg grate temp (°C)	Avg exhaust gas temp (°C)	Avg amount dry exhaust gas (m^3/hr)	Avg O_2 conc (%)	Avg CO_2 conc (%)	Avg CO conc (ppm)	Cl ion conc (mg/m^3 N)	References
NaCl-I NP B	0.99	578	594	461	225	15.0	4.5	251	44	Yasuhara et al. 2002
NaCl/$CuCl_2$-I NP	1.13	662	675	473	240	14.7	4.9	350	200	Yasuhara et al. 2005b
NaCl/$MgCl_2$-I NP	1.34	689	595	456	222	15.9	4.0	400	120	Yasuhara et al. 2005b
NaCl/$MnCl_2$-I NP	1.06	710	664	515	217	15.6	4.2	300	240	Yasuhara et al. 2005b
NaCl/$FeCl_2$-I NP	1.14	659	660	481	206	14.6	5.0	310	170	Yasuhara et al. 2005b
NaCl/$NiCl_2$-I NP	1.14	705	645	517	218	15.6	4.2	400	250	Yasuhara et al. 2005b
NaCl/$CoCl_2$-I NP	1.07	664	673	476	226	15.2	4.4	350	120	Yasuhara et al. 2005b
NaCl-I NP + fly ash	1.36	512	535	—	227	16.3	3.2	550	176	Yasuhara et al. 2005c

Table 3. Combustion Chamber Conditions and Concentrations of O_2, CO, CO_2, and Cl Ion in Exhaust Gases from Incinerated Plastics and Plastics + NaCl.

Combustion sample	Chlorine cont (wt%)	Avg chamber temp (°C)	Avg grate temp (°C)	Avg exhaust gas temp (°C)	Avg amount dry exhaust gas (m^3/hr)	Avg O_2 conc (%)	Avg CO_2 conc (%)	Avg CO conc (ppm)	Cl ion conc (mg/m^3 N)	References
Polyethylene (PE)	<0.0005	656	799	533	231	14.5	4.6	77	4	Katami et al. 2002
Polystyrene (PS)	<0.0005	738	749	514	263	13.9	5.5	130	2	Katami et al. 2002
PS foam	0.025	668	951	513	274	15.2	4.0	30.9	—	Yasuhara et al. 2005a
Polyethylene tetraphthalate (PET)	ND	656	769	550	240	13.7	5.2	14	5	Katami et al. 2002
PE + NaCl	3.14	759	722	502	279	14.2	4.8	118	41	Yasuhara et al. 2002
PS + NaCl	2.38	803	810	570	265	13.3	6.0	150	60	Yasuhara et al. 2002
PET + NaCl	2.80	672	932	589	232	13.3	5.5	3.0	38	Yasuhara et al. 2002
NaCl-I NP + PE	—	473	—	434	1,020	17.2	3.2	990	59	Yasuhara et al. 2001

Table 4. Combustion Chamber Conditions and Concentrations of O_2, CO, CO_2, and Cl Ion in Exhaust Gases from Incinerated PVC and PVC with Other Plastics (PE, PS, and PET) and Newspaper.

Combustion sample	Chlorine cont (wt%)	Avg chamber temp (°C)	Avg grate temp (°C)	Avg exhaust gas temp (°C)	Avg amount dry exhaust gas (m^3/hr)	Avg O_2 conc (%)	Avg CO_2 conc (%)	Avg CO conc (ppm)	Cl ion conc (mg/m^3 N)	References
PE + polyvinylchloride (PVC)	4.58	747	771	576	255	12.7	5.8	41	420	Katami et al. 2002
PS + PVC	4.58	677	731	570	242	14.2	4.7	40	170	Katami et al. 2002
PET + PVC	4.63	689	891	600	230	13.4	5.5	2	92	Katami et al. 2002
PVC in low CO conc	51.3	633	900	542	298	13.6	5.2	42	1,600	Katami et al. 2002
PVC in in high CO conc	51.3	531	742	448	267	16.5	3.2	880	1,800	Katami et al. 2002
PVC-I NP	4.38	699	637	—	220	14.5	4.2	280	470	Yasuhara et al. 2005c
NP + PVC	35.7	456	—	416	967	17.4	3.0	1,500	640	Yasuhara et al. 2001
$NaHCO_3$-I NP	0.02	741	638	—	231	13.3	5.5	49	3.5	Yasuhara et al. 2005c
$NaHCO_3$/PVC-I NP	4.51	726	619	—	220	15.2	5.3	310	560	Yasuhara et al. 2005c

Table 5. Combustion Chamber Conditions and Concentrations of O_2, CO, CO_2, and Cl Ion in Exhaust Gases from Incineration of Various Kinds of Wood and Leaves.

Combustion sample	Chlorine cont (wt%)	Avg chamber temp (°C)	Avg grate temp (°C)	Avg exhaust gas temp (°C)	Avg amount dry exhaust gas (m^3/hr)	Avg O_2 conc (%)	Avg CO_2 conc (%)	Avg CO conc (ppm)	Cl ion conc (mg/m^3 N)	References
London plane tree branches	—	456	—	435	907	16.5	3.7	1,090	ND	Yasuhara et al. 2001
Bleached kraft pulp (BKP)	<0.01	702	647	525	204	15.5	4.4	77	1.2	Yasuhara et al. 2002
Pine wood (PW)	<0.01	563	597	472	228	15.7	4.0	310	3.2	Yasuhara et al. 2003
Japanese cedar wood (CW)	0.02	537	544	444	240	15.9	3.9	730	6.5	Yasuhara et al. 2003
Beech wood	<0.01	917	969	667	242	10.9	7.1	<5.0	2.9	Yasuhara et al. 2003
Leaves from cherry tree	0.0074	746	705	408	219	15.4	4.2	240	15	Katami et al. 2004a
Leaves from konara oak tree	0.023	692	635	513	237	16.0	5.5	270	15	Katami et al. 2004a

Table 6. Combustion Chamber Conditions and Concentrations of O_2, CO, CO_2, and Cl Ion in Exhaust Gases from Incineration of Various Kinds of Wood Impregnated with Seawater, Chlordane, or Pentachlorophenol.

Combustion sample	Chlorine cont (wt%)	Avg chamber temp (°C)	Avg grate temp (°C)	Avg exhaust gas temp (°C)	Avg amount dry exhaust gas (m^3/hr)	Avg O_2 conc (%)	Avg CO_2 conc (%)	Avg CO conc (ppm)	Cl ion conc (mg/m^3 N)	References
NaCl-I BKP	4.25	684	594	504	209	16.2	3.8	159	46	Yasuhara et al. 2002
Sea water (SW)-I PW	1.8	536	514	491	218	15.9	3.7	670	120	Yasuhara et al. 2003
SW-I CW with bark	1.3	575	558	469	249	16.0	3.8	560	92	Yasuhara et al. 2003
SW-I CW without bark	0.73	567	576	477	220	16.0	3.7	330	34	Yasuhara et al. 2003
Chlordane-I waste wood	0.02	805	866	615	237	12.7	6.0	5.6	4.0	Yasuhara et al. 2003
Pentachlorophenol-I waste wood	0.34	842	901	614	226	12.2	6.4	5.0	20	Yasuhara et al. 2003

Table 7. Combustion Chamber Conditions and Concentrations of O_2, CO, CO_2, and Cl Ion in Exhaust Gases from Incineration of Miscellaneous Samples.

Combustion sample	Chlorine cont (wt%)	Avg chamber temp (°C)	Avg grate temp (°C)	Avg exhaust gas temp (°C)	Avg amount dry exhaust gas (m^3/hr)	Avg O_2 conc (%)	Avg CO_2 conc (%)	Avg CO conc (ppm)	Cl ion conc (mg/m^3 N)	References
Plastic cover for air conditioner	0.036	857	1,076	617	231	10.8	7.1	7.4	—	Yasuhara et al. 2005a
Cardboard for packaging	0.027	760	754	—	223	13.0	5.9	<5	0.3	Yasuhara et al. 2005c
Lightweight coated papers	0.01	682	721	—	227	14.8	4.5	140	8.0	Yasuhara et al. 2005c
No printing (NP)-NP	0.0064	709	726	504	265	13.5	5.9	15.2	75	Yasuhara et al. 2006
NP-NP + PVC	4.59	674	697	570	241	15.4	4.4	103	1,400	Yasuhara et al. 2006
Food sample A	0.99	652	753	478	247	14.1	5.5	18	65	Katami et al. 2004b
Food sample B	1.22	696	756	527	245	13.9	5.2	22	65	Katami et al. 2004b
PVC-coated electric wire	10.6	787	1,029	646	214	12.2	5.9	12	480	Yasuhara et al. 2005b
PVC-removed electric wire	17.1	763	1,010	639	223	12.6	5.8	13	490	Yasuhara et al. 2005b

according to the combustion conditions because it may depend significantly on the nature of the samples. However, it is possible to adjust CO concentrations with chamber and grate temperatures so long as the same materials are incinerated. For example, the CO concentration of PVC samples was adjusted to a high level (880 ppm) and a low level (42 ppm) to investigate the formation difference of dioxins (Katami et al. 2002). Cl ion concentrations were somewhat proportional to the chloride content of the samples. When polyethylene (PE, chloride content = <0.0005%) was incinerated, Cl ion concentration in the exhaust gas was 4 mg/m^{3}N, in contrast to 41 mg/m^{3}N in the exhaust gas from the combustion of PE + NaCl (chloride content = 3.14%) (Katami et al. 2002). It should be noted that the Cl ion concentration tends to vary over the quantity samples incinerated/hr. Therefore, the factor influencing Cl ion concentration in an exhaust gas is not only the chloride content in the samples.

III. Exhaust Gas Collection

Exhaust gases were collected at the sampling port located between the combustion chamber and the cyclone (see Fig. 1). The apparatus used for the collection of exhaust gases from the incinerator is shown in Fig. 2. Dust in the exhaust gas was removed with an in-line silica-fiber thimble filter. Exhaust gases were next drawn into two 1-L impingers, each containing 150 mL water (washed with hexane before use) and an empty impinger connected in series. The empty impinger was further connected to a column packed with 40 g XAD-2 resin that was interfaced to a 1-L impinger containing 250 mL diethylene glycol and an empty impinger connected in series. The impingers were kept at 5°C during sample collections in an ice-cooled water bath. Exhaust gas was drawn using a diaphragm vacuum pump with a flow rate of 21–23 L/min, which was the same as that of exhaust gas in the duct. Dioxin standards (0.5 ng each), $^{13}C_{12}$-1,2,3,4-T_4CDD, $^{13}C_{12}$-1,2,3,4,7,8-H_6CDF, and 1,2,3,4,7,8,9-H_7CDF, were added here to the first impinger for sampling-spike recovery tests before sampling.

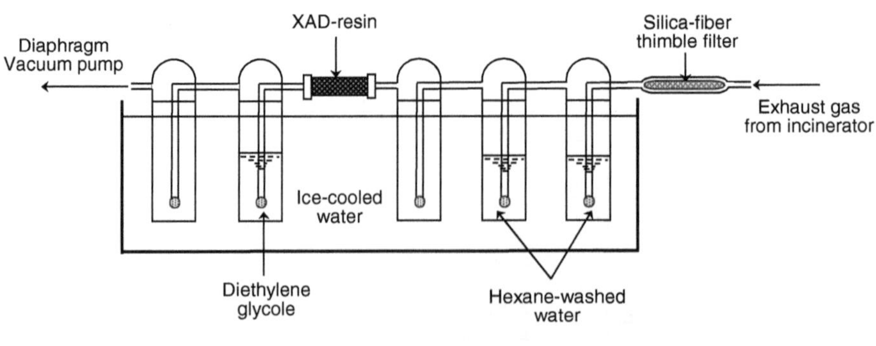

Fig. 2. Apparatus used for collection of incinerator exhaust gases.

After sampling, the dioxin standards, $^{13}C_{12}$-2,3,7,8-T_4-CDD, 1,2,3,7,8-P_5CDD, 1,2,3,6,7,8-H_6CDD, 1,2,3,4,6,7,8-H_7CDD, 1,2,3,4,6,7,8,9-O_8CDD, $^{13}C_{12}$-2,3,7,8-T_4CDF, 1,2,3,7,8-P_5CDF, 1,2,3,4,7,8-H_6CDF, 1,2,3,4,6,7,8-H_7CDF, 1,2,3,4,6,7,8,9-O_8CDF, $^{13}C_{12}$-3,3′,3,4′-T_4CB, 3,4,4′,5T_4CB, 3,3′,4,4′,5-P_5CB, 2′,3,4,4′,5-P_5CB, 3,3′,4,4′,5,5′-H_6CB, 2,3′,4,4′,5,5′-H_6CB, and 2,3,3′,4,4′,5,5′-H_7CB, were spiked (0.5 ng each) at this point in the dust in the silica-fiber thimble filter and the diethylene glycol solution in the impinger for extraction and cleanup recovery tests. The dust trapped in the silica-fiber thimble filter (~1 g) was washed with 20 mL HCL solution (2 mol/L) and combined with the XAD-2 resin after drying. The resin was then extracted with 200 mL toluene using a Soxhlet extractor for 16 hr. The HCL solution from the dust was triple extracted with 10-mL portions of dichloromethane. The water and diethylene glycol (250 mL) in the impingers, as well as the water trapped (trace) in the empty impingers, were combined and triple extracted with dichloromethane (1/10 volume of aqueous solution). After all extracts were combined, the extract was condensed by distillation using a rotary flash evaporator, and the combined samples were cleaned with multilayer silicagel chromatography (JMHW 1997). The sample was further cleaned with a 120-mL hexane/dichloromethane (1/1) solution using alumina column chromatography. After an extract from each sample was condensed using a rotary flash evaporator, 0.5 ng each of internal dioxin standard ($^{13}C_{12}$-1,3,6,8-T_4CDD and 1,2,3,7,8,9-H_6CDD) was added for quantitative analysis of dioxins by GC/MS. The volume of the sample was adjusted to exactly 50 mL with *n*-nonane. The samples were subsequently analyzed by GC/MS for dioxins.

IV. Dioxin Analysis in Exhaust

A typical and representative method for dioxin analysis was previously reported (Katami et al. 2004b). An Agilent model 5,890 gas chromatograph (GC) interfaced to a Micromass double focus MS (Auto Spec; ULTIMA, Manchester, England) was used for determination of PCDDs, PCDFs, and coplanar PCBs. Gas chromatographic columns were a 60 m × 0.32 mm i.d. (d_f = 0.2 μm) SP-2331 bonded-phase fused-silica capillary column (Supelco, Bellefonte, PA, USA) for Cl_1–Cl_6-PCDDs and Cl_1–Cl_6–PCDFs; a 30 m × 0.25 mm i.d. (d_f = 0.25 μm) DB-5 bonded-phase fused-silica capillary column (J & W Scientific, Folsom, CA, USA) for $Cl_{7,8}$-PCDDs and $Cl_{7,8}$–PCDFs; or a 60 m × 0.32 mm i.d. (d_f = 0.25 μm) DB-5 bonded-phase fused-silica capillary column (J & W Scientific) for coplanar PCBs. Gas chromatographic oven temperatures were programmed from 130° to 190°C at 20°C/min and then to 250°C at 2°C/min for the SP-2331 column; programmed from 130° to 280°C at 10°C/min for the 30 m-DB-5 column; and programmed from 150°C (held for 1 min) to 185°C at 20°C/min and then programmed at 2° to 245°C (held for 3 min), and at 6° to 290°C (held for 10 min) for a 60 m-DB-5 column. The linear velocity of the helium carrier gas was 30 cm/sec. The

Table 8. Recovery Efficiencies of Standard Dioxins.

Cleanup-spike		Sampling-spike	
Dioxins	Recovery (%)	Dioxins	Recovery (%)
$^{13}C_{12}$-2, 3, 7, 8-T_4CDD	81 ± 12.8	$^{13}C_{12}$-1, 2, 3, 4-T_4CDD	88 ± 13.2
$^{13}C_{12}$-1, 2, 3, 7, 8-P_5CDD	89 ± 12.4	$^{13}C_{12}$-1, 2, 3, 4, 7, 8-H_6CDD	87 ± 8.3
$^{13}C_{12}$-1, 2, 3, 6, 7, 8-H_6CDD	87 ± 10.8	$^{13}C_{12}$-1, 2, 3, 4, 6, 7, 8-H_7CDF	94 ± 13.1
$^{13}C_{12}$-1, 2, 3, 4, 6, 7, 8-H_7CDD	85 ± 10.8		
$^{13}C_{12}$-1, 2, 3, 4, 6, 7, 8, 9-O_8CDD	79 ± 5.3		
$^{13}C_{12}$-2, 3, 7, 8-T_4CDF	80 ± 12.0		
$^{13}C_{12}$-1, 2, 3, 7, 8-P_5CDF	89 ± 10.9		
$^{13}C_{12}$-1, 2, 3, 4, 7, 8-H_6CDF	88 ± 8.5		
$^{13}C_{12}$-1, 2, 3, 4, 7, 8, 9-H_7CDF	79 ± 8.5		
$^{13}C_{12}$-1, 2, 3, 4, 6, 7, 8, 9-O_8CDF	77 ± 4.5		
$^{13}C_{12}$-3, 3′4, 4′-T_4CB	79 ± 16.8		
$^{13}C_{12}$-3, 4, 4′, 5-T_4CB	80 ± 16.2		
$^{13}C_{12}$-3, 3′, 4, 4′, 5-P_5CB	79 ± 12.7		
$^{13}C_{12}$-2′, 3, 4, 4′, 5-P_5CB	82 ± 5.3		
$^{13}C_{12}$-3, 3′, 4, 4′, 5, 5′-H_6CB	89 ± 13.5		
$^{13}C_{12}$-2, 3′, 4, 4′, 5, 5′-H_6CB	82 ± 6.1		
$^{13}C_{12}$-2, 3, 3′, 4, 4′, 5, 5′-H_7CB	93 ± 9.5		

injector temperatures were 250°C for the SP-2331 column and 280°C for the DB-5 columns. MS ion source temperatures were 250°C for the SP-2331 column and 289°C for the DB-5 column. MS ionization voltage was 35 eV.

A 60-m SP-2331 bonded-phase column has been used most widely to analyze specific dioxin isomers in various samples, such as papermill wastewater (Tiernan et al. 1989), soft-shell clam (Brown et al. 1994), sediments (Tong et al. 1990), exhaust gases from car (Bingham et al. 1989), and exhaust gases from an incinerator (Oehme and Kirschmer 1984; Katami et al. 2004b). A 30-m DB-5 bonded-phase column is also commonly used for determination of total dioxins and/or separation of PCDDs and PCDFs (Marquis et al. 1995; Wiliams and Giesy 1995; Kleopfer et al. 1989). For example, a 30-m DB-5 was used for the primary analysis of fish samples, and then a 60-m SP2331 was used to verify results (Marquis et al. 1994).

One of the most widely used GC/MS quantitation methods for dioxins in different kinds of samples is the use of labeled internal standards, $^{13}C_{12}$-PCDDs/Fs and $^{13}C_{12}$-coplanar PCBs. For example, in the case of 2,3,7,8-T_4CDD, its concentration is determined by comparing the ratio of the m/z at 322 from unlabeled 2,3,7,8-T_4CDD and m/z 334 from labeled 2,3,7,8-T_4CDD to the calibration curve, which is established using the standard solutions (Rappe and Buser 1980).

Typical recovery efficiencies of typical standard $^{13}C_{12}$-dioxins with sampling spike and cleanup spike are shown in Table 8. Values are mean ± standard deviation ($n = 11$). The recoveries in both the sampling spike and the cleanup spike were satisfactory values greater than 77% (Yasuhara et al. 2002).

V. Dioxins Found in Exhaust

The term *dioxins* is commonly used to describe a class of chlorinated aromatic hydrocarbons, including PCDDs, PCDFs, and coplanar PCBs. The general structures of the 75 PCDDs, 135 PCDFs, and 209 PCBs are shown in Fig. 3. Table 9 shows the dioxins measured in exhaust gases from an incinerator. They are mono- (M_1CDD), di- (D_2CDD), tri- (T_3CDD), tetra- (T_4CDD), penta- (P_5CDD), hexa- (H_6CDD), hepta- (H_7CDD), and octa-chlorodibenzo-*p*-dioxin (O_8CDD); mono- (M_1CDF), di- (D_2CDF), tri- (T_3CDF), tetra- (T_4CDF), penta- (P_5CDF), hexa- (H_6CDF), hepta- (H_7CDF), and octa-chlorodibenzofuran (O_8CDF); and tetra- (T_4CB), penta- (P_5CB), hexa- (H_6CB), and hepta-chloro-coplanar biphenyl (H_7CB). The most biologically active PCB congeners are 3,4,4′,5-tetra-, 3,3′,4,4′-tetra-, 3,3′,4,4′,5-penta-, and 3,3′,4,4′,5,5′-hexachlorobiphenyl, and their toxicities have been widely reported (Safe et al. 1985).

A. Newspaper Impregnated with NaCl, KCl, or $CaCl_2$

Old newspapers comprise the largest part of domestic wastes. They have been collected separately from other domestic garbage together with other

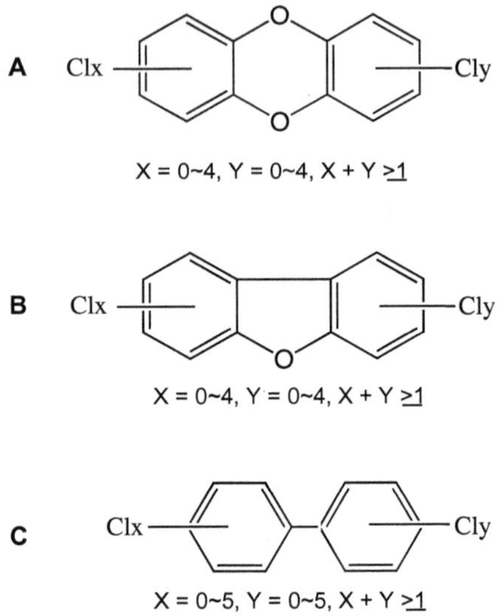

Fig. 3. General structures of (A) 75 polychlorinated dibenzo-*p*-dioxins (PCDDs), (B) 135 polychlorinated dibenzo furans (PCDFs), and (C) 209 polychlorinated biphenyls (PCBs).

sorts of paper for the purpose of recycling. However, they are eventually burned in an incinerator. The chlorine content of the newspapers combusted in the experiments was 0.0064%. Metal contents in the newspapers used in the present study were 0.33 μg/g Cr, 2.7 μg/g Mn, 33 μg/g Fe, 0.23 μg/g Ni, 8.6 μg/g Cu, 2.3 μg/g Zn, 0.01 μg/g Cd, and 0.75 μg/g Pd; As (arsenic) was not detected. The calculated amounts of Cu and Fe present in the incinerator during combustion of 50 kg newspapers were 0.43 g and 1.65 g, respectively, suggesting that the low concentrations of these metals did not significantly contribute to dioxin formation.

Figure 4 shows the results of dioxin analysis in the exhaust gases from the combustion of newspapers impregnated with NaCl, KCl, or $CaCl_2$ in an incinerator. The figure was constructed with the results from two articles (Yasuhara et al. 2001, 2002). The notation "×10" indicates that the actual values are 10 times the values shown in the figure. Table 1 shows the combustion chamber conditions and concentrations of O_2, CO, CO_2, and Cl ion in the exhaust gases along with chlorine content of the samples shown in Fig. 4. When NaCl-I NP with different chlorine contents—B (0.99%), C (1.88%), and D (4.08%)—were combusted under similar chamber and grate temperatures, the higher the chlorine content, the higher the amounts of dioxins that were formed. The decreasing order of total dioxins formed from these samples was D (174 ng/g) > C (121 ng/g) > B (49 ng/g). NaCl-I NP E contained exactly the same amount of as NaCl-I NP D, but it was combusted at a higher temperature (791°C) than NaCl-I NP (751°C).

Table 9. Dioxin Congeners Determined in Exhaust Gases from Incineration of Various Samples.

M$_1$CDD	1, 2, 8, 9-	2, 4-	1, 4, 7, 8-	1, 2, 4, 8, 9-
2-	1, 2, 8, 9-	3, 7-	1, 3, 6, 9-	2, 3, 4, 7, 8-
1-		2, 7-	1, 2, 3, 7-	1, 2, 3, 8, 9-
	P$_5$CDD	2, 3-	1, 6, 7, 8-	2, 3, 4, 6, 7-
D$_2$CDD	1, 2, 4, 6, 8-	3, 6-	1, 2, 3, 4-	
1, 3-	1, 2, 4, 7, 9-	2, 8-	2, 4, 6, 8-	H$_6$CDF
2, 3-	1, 2, 3, 6, 8-	2, 6-	1, 2, 3, 8-	1, 2, 3, 4, 6, 8-
2, 7-	1, 2, 4, 7, 8-	1, 9-	1, 4, 6, 7-	1, 3, 4, 6, 7, 8-
2, 8-	1, 2, 3, 7, 9-	3, 4-	1, 2, 3, 6-	1, 3, 4, 6, 7, 9-
1, 4-	1, 2, 4, 6, 9-	4, 6-	1, 3, 4, 9-	1, 2, 4, 6, 7, 8-
1, 7-	1, 2, 3, 4, 7-		1, 2, 7, 8-	1, 2, 4, 6, 7, 9-
1, 8-	1, 2, 3, 7, 8-	T$_3$CDF	1, 2, 6, 7-	1, 2, 3, 4, 7, 8-
1, 6-	1, 2, 3, 6, 9-	1, 3, 7-	1, 2, 7, 9-	1, 2, 3, 4, 7, 9-
1, 2-	1, 2, 4, 6, 7-	1, 3, 8-	1, 4, 6, 9-	1, 2, 3, 6, 7, 8-
1, 9-	1, 2, 4, 8, 9-	1, 3, 6-	1, 2, 4, 9-	1, 2, 4, 6, 8, 9-
	1, 2, 3, 4, 6-	1, 3, 4-	2, 3, 6, 8-	1, 2, 3, 4, 6, 7-
T$_3$CDD	1, 2, 3, 6, 7-	1, 6, 8-	2, 4, 6, 7-	1, 2, 3, 6, 7, 9-
1, 3, 7-	1, 2, 3, 8, 9-	1, 2, 4-	1, 2, 3, 9-	1, 2, 3, 4, 6, 9-
1, 3, 8-		1, 4, 7-	2, 3, 4, 7-	1, 2, 3, 6, 8, 9-
1, 3, 6-	H$_6$CDD	1, 6, 7-	1, 2, 6, 9-	1, 2, 3, 7, 8, 9-
1, 2, 4-	1, 2, 3, 4, 6, 8-	1, 7, 8-	2, 3, 7, 8-	1, 2, 3, 4, 8, 9-
1, 3, 9-	1, 2, 4, 6, 7, 9-	1, 4, 8-	2, 3, 4, 8-	2, 3, 4, 6, 7, 8-
1, 2, 3-	1, 2, 4, 6, 8, 9-	1, 2, 3-	2, 3, 4, 6-	
1, 7.8-	1, 2, 3, 6, 7, 9-	1, 4, 6-	2, 3, 6, 7-	H$_7$CDF
1, 2, 7-	1, 2, 3, 6, 8, 9-	1, 2, 7-	3, 4, 6, 7-	1, 2, 3, 4, 6, 7, 8-
1, 2, 8-	1, 2, 3, 4, 7, 8-	2, 4, 7-	1, 2, 8, 9-	1, 2, 3, 4, 6, 7, 9-
1, 4, 6-	1, 2, 3, 6, 7, 8-	1, 2, 8-		1, 2, 3, 4, 6, 8, 9-
1, 2, 9-	1, 2, 3, 4, 6, 9-	1, 2, 6-	P$_5$CDF	1, 2, 3, 4, 7, 8, 9-
2, 3, 7-	1, 2, 3, 7, 8, 9-	2, 4, 8-	1, 3, 4, 6, 8-	
1, 4, 7-	1, 2, 3, 4, 6, 7-	2, 4, 6-	1, 2, 4, 6, 8-	O$_8$CDF
		2, 3, 7-	1, 3, 6, 7, 8-	1, 2, 3, 4, 6, 7, 8, 9-
T$_4$CDD	H$_7$CDD	1, 4, 9-	1, 3, 4, 7, 9-	
1, 3, 6, 8-	1, 2, 3, 4, 6, 7, 9-	2, 3, 4-	1, 2, 3, 6, 8-	
1, 3, 7, 9-	1, 2, 3, 4, 6, 7, 8-	2, 3, 8-	1, 3, 4, 7, 8-	T$_4$CB
1, 3, 7, 8-		3, 4, 7-	1, 2, 4, 7, 8-	3, 4, 4′, 5-
1, 3, 6, 9-	O$_8$CDD	2, 6, 7-	1, 2, 4, 7, 9-	3, 3′, 4, 4′-
1, 2, 4, 8-	1, 2, 3, 4, 6, 7, 8, 9-	1, 2, 9-	1, 3, 4, 6, 7-	
1, 2, 6, 8-		3.4, 6-	1, 2, 4, 6, 7-	P$_5$CB
1, 4, 7, 8-			1, 4, 6, 7, 8-	3, 3′, 4, 4′, 5-
2, 3, 7, 8-	M$_1$CDF	T$_4$CDF	1, 2, 3, 4, 7-	2′, 3, 4, 4′, 5-
1, 2, 3, 7-	1-	1, 3, 6, 8-	1, 3, 4, 6, 9-	2, 3′, 4, 4′, 5-
1, 2, 3, 4-	2-	1, 3, 7, 8-	1, 2, 3, 4, 8-	2, 3, 3′, 4, 4′-
1, 2, 4, 9-	3-	1, 3, 7, 9-	1, 2, 3, 7, 8-	2, 3, 4, 4′, 5-
1, 2, 4, 6-	4-	1, 3, 4, 7-	1, 2, 3, 4, 6-	
1, 2, 3, 8-		1, 4, 6, 8-	1, 2, 3, 7, 9-	H$_6$CB
1, 2, 3, 6-	D$_2$CDF	1, 2, 4, 7-	1, 2, 3, 6, 7-	3, 3′, 4, 4′, 5, 5′-
1, 2, 7, 9-	1, 3-	1, 3, 6, 7-	1, 2, 4, 6, 9-	2, 3′, 4, 4′, 5, 5′-
1, 4, 6, 9-	1, 7-	1, 3, 4, 8-	1, 2, 6, 7, 8-	2, 3, 3′, 4, 4′, 5-
1, 2, 7, 8-	1, 4-	1, 3, 4, 6-	1, 2, 6, 7, 9-	2, 3, 3′, 4, 4′, 5′-
1, 2, 3, 9-	1, 8-	1, 2, 4, 8-	1, 2, 3, 6, 9-	
1, 2, 6, 9-	1, 6-	1, 2, 4, 6-	2, 3, 4, 6, 8-	H$_7$CB
1, 2, 6, 7-	1, 2-	1, 2, 6, 8-	1, 2, 3, 4, 9-	2, 3, 3′, 4, 4′, 5, 5′

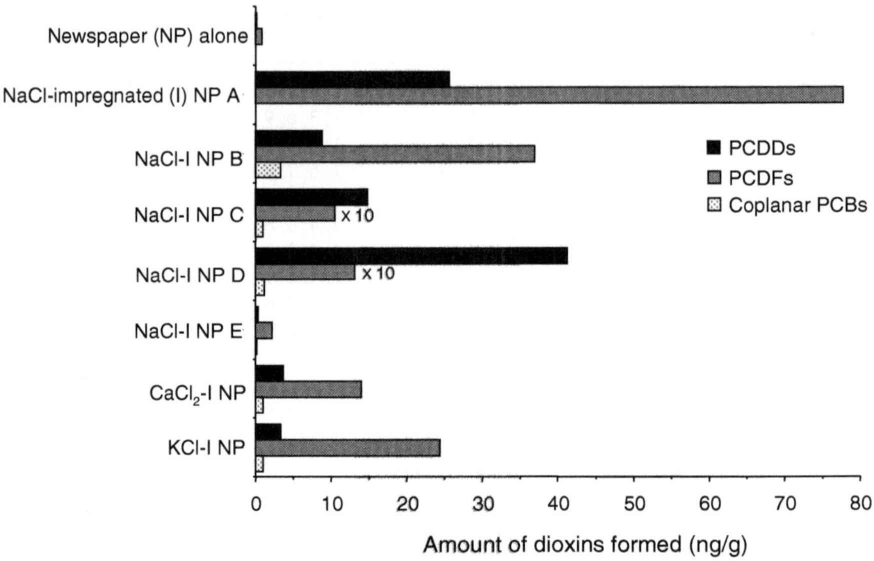

Fig. 4. Dioxin analysis of exhaust gases from combustion of newspapers impregnated with NaCl, KCl, or $CaCl_2$.

Consequently, NaCl-I NP E produced much less dioxins (2.68 ng/g) than NaCl-I NP D did (174 ng/g), suggesting that dioxin formation could be minimized by appropriate combustion conditions, even when the samples have a high chlorine content.

NaCl-I NP A (chlorine content = 3.1%) was combusted in a different experiment from the other three samples (B, C, and D). Therefore, it is difficult to compare its results to those of others, but the total amount of dioxins formed was within the comparable range (103 ng/g). The other research group also reported dioxin formation from NP alone (21.6 pg/g) and NaCl-I NP (51.9 pg/g) upon combustion (Katsumata et al. 2003).

Dioxin formation from alkaline metal chlorides-I NP, in which chlorine contents were similar to each other, was 18.6 ng/g from $CaCl_2$-I NP, 28.6 ng/g from KCl, and 49.0 ng/g from NaCl. These formation differences may be due to the bond energy differences between the metal and the Cl atom (455 kj/mol for $CaCl_2$, 422 kj/mol for KCl, and 407 kj/mol for NaCl). The lesser the bond energy, the easier it is to release the Cl ion, and consequently more dioxins are produced.

Formation of total PCDFs was much higher than that of PCDDs in all eight samples. The total PCDFs composed 78%–92% of the total dioxin formed. Coplanar PCBs were found at the lowest level among the dioxins formed. Their formation levels ranged from 0 (NP alone) to 3.32 ng/g (NaCl-I NP B).

B. Newspaper Impregnated with NaCl, $FeCl_3$, $MnCl_2$, and Other Salts

Dioxin formation in the presence of metals, such as Cu and Fe, at low temperatures (200°–500°C) has been reported (Tuppurainen et al. 1998). For example, it was hypothesized that the *de novo* synthesis at 300°C from particulate carbon proceeds through a direct ligand transfer of the halide, in which metal ions (Cu, Fe) are involved (Stieglitz et al. 2003). However, there are only a few reports on dioxin formation in a combustion chamber of incinerators in the presence of metals at high temperatures (>500°C), and knowledge of the formation mechanisms of dioxins in high-temperature processes is still considerably limited. More experiments and theoretical studies to elucidate the details of dioxin formation during combustion are a pressing need.

Figure 5 shows the results of dioxin analysis of the exhaust gases from the combustion of newspapers impregnated with NaCl and various inorganic metal chlorides in an incinerator. This figure was constructed with the results from one article (Yasuhara et al. 2005b). The "×10" indicates that the actual values are 10 times the values shown in the figure. Table 2 shows the combustion chamber conditions and concentrations of O_2, CO, CO_2, and Cl ion in the exhaust gases, along with the chlorine content, of the samples shown in Fig. 5. Newspaper (approximately 5 g) was soaked in a solution containing NaCl and each different metal chloride for 20 min and then dried in an oven at 110°C (Yasuhara et al. 2005b).

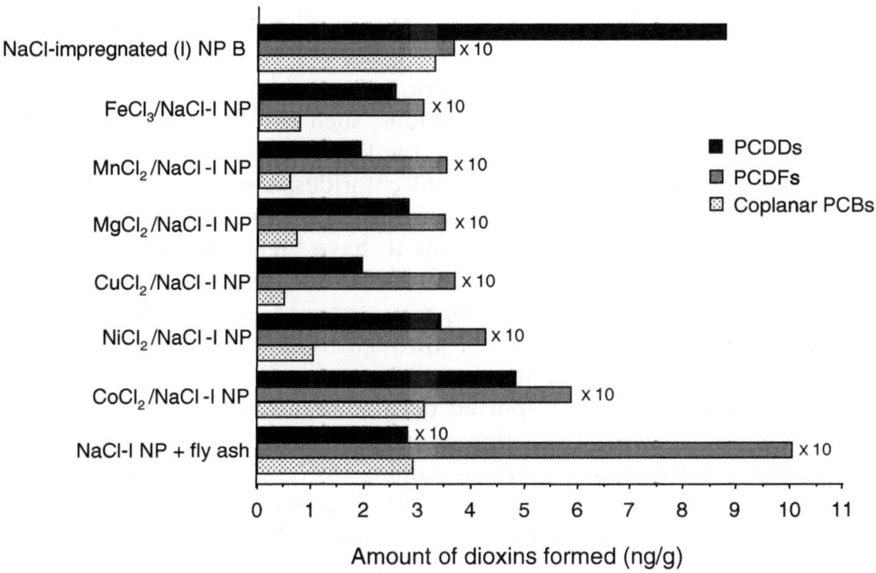

Fig. 5. Dioxin analysis of exhaust gases from combustion of newspapers impregnated with NaCl and various inorganic metal chlorides.

When newspaper alone was combusted, the total amount of dioxins (total of PCDDs, PCDFs, and coplanar PCBs) reached only trace level (0.18 ng/g). This value is quite low compared with those of the samples with chloride. NaCl-impregnated (I) newspaper (NP) alone yielded the greatest amount of total dioxins (103 ng/g) among the samples shown in this figure. The highest level of PCDDs (8.79 ng/g) was also formed from NaCl-I NP alone, whereas the highest level of PCDFs (59.0 ng/g) was obtained from $CuCl_2$/NaCl-impregnated newspaper. PCDFs comprised more than 80% of the total dioxins formed from these samples. The results show that the addition of metal chlorides decreased the formation of dioxins from NaCl-I NP upon combustion. The decreasing order of total dioxins formed from NaCl-I NP with other metal chlorides is $CuCl_2$ (67.0 ng/g) > $MgCl_2$ (47.2 ng/g) > $MnCl_2$ (39.7 ng/g) > $FeCl_3$ (38.7 ng/g) > $NiCl_2$ (37.8 ng/g) > $CoCl_2$ (34.2 ng/g). The dioxin formation from NaCl-I NP upon combustion seemed to be reduced by the additional impregnation of other metals. Addition of fly ash to NaCl-I NP produced 139 ng/g total dioxins upon combustion under similar conditions (Yasuhara et al. 2005c). Inorganic elements in the fly ash comprised (wt%) Ca (49.0), Cl (28.0), K (8.5), Si (3.3), S (3.1), Al (2.2), Zn (1.9), Fe (1.0), Ti (0.97), Na (0.88), Pb (0.39), Cd (0.23), and Cu (0.15). Ca comprised nearly 50% of the fly ash. Only trace amounts of heavy metals, which are known to catalyze dioxin formation, were present in the fly ash. Experiments in a laboratory-scale fluidized-bed reactor using a synthetic fuel that simulated municipal solid waste showed that small fly ash particles (<2.2 μm) had a significant influence on the formation of chlorinated aromatic compounds (Fängmark et al. 1994). Also, addition of dioxin-free fly ash to NP, PE, and PVC increased dioxin formation significantly (Takasugi et al. 2000).

Dioxin formation from heated chlorophenols was increased by metals, such as Cu and Fe, and by metal oxides, such as CuO, NiO, and Fe_2O_3 (Gullett et al. 1990, 1992). When unburned carbonaceous particles, such as soot, were burned with various metal chlorides, the increasing order of metal activity toward dioxin formation was KCl < $CaCl_2$ < $FeCl_3$ << $CuCl_2$ (Kuzuhara et al. 2003). Copper seems to have higher activity than other metals tested; it acted not only as a catalyst but also as a transmitter of heterogeneous chlorine. Excess copper present as $CuCl_2$ in municipal solid waste chlorinated activated carbon through to chlorinated benzenes and PCDD/F at 300°C (Jay and Stieglitz 1991).

A correlation has been reported between chlorine content and dioxin formation if the level of chlorine content is higher than 1% (Wikström et al. 1996). When newspapers containing 0.99%, 1.88%, and 4.08% of chlorine were burned, 49.0, 121, and 174 ng/g of total dioxins were formed, respectively, indicating that there is an obvious relationship between chlorine content and dioxin formation (Yasuhara et al. 2002).

The mechanisms of dioxin formation and degradation under the presence of metals are still not completely understood. Copper compounds, including CuO, have been known to catalyze dioxin formation, but then also catalyze the decomposition of dioxins, depending on their concentration (Luijk et al. 1994). In addition to copper, various metals are also known to play a role in dioxin formation (Addink and Olie 1995). Some metals also catalyze oxidative degradation of organic compounds, and their decreasing order of activity is Cu > Co > Fe > Ni > Pt (Inui et al. 1982). Therefore, it is hypothesized that the metals added to the combustion materials acted as catalysts for the oxidative degradation of organic compounds during combustion and consequently reduced dioxin formation in the exhaust gases.

The foregoing reports indicate that some activities of alkaline metals toward dioxin formation remain incompletely understood. More intensive study on the role of metals in dioxin formation from incinerators used for burning domestic and industrial wastes is in order.

C. Plastics Alone and with NaCl

Polyethylene (PE), polystyrene (PS), and polyethylene terephthalate (PET) have been used to produce plastic sheets and bags, which are widely used at grocery stores and individual homes. Eventually, these plastic materials are combusted as domestic wastes in an incinerator in the presence of chlorines.

Figure 6 shows the results of dioxin analysis in the exhaust gases from the combustion of plastics and plastics + NaCl in an incinerator. This figure

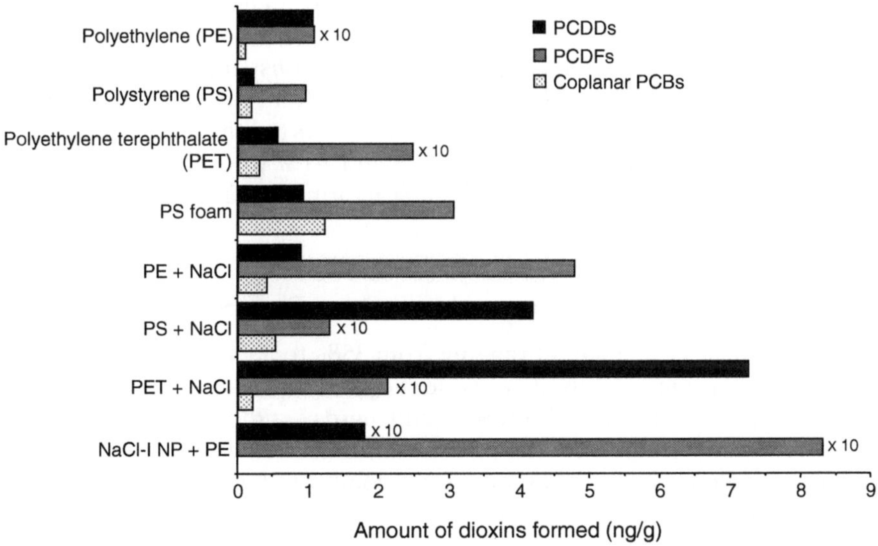

Fig. 6. Dioxin analysis of exhaust gases from combustion of plastics and plastics + NaCl.

was constructed from the results of three articles (Yasuhara et al. 2002, 2005a; Katami et al. 2002). The "×10" indicates that the actual values are 10 times the values shown in the figure. Table 3 shows the combustion chamber conditions and concentrations of O_2, CO, CO_2, and Cl ion in the exhaust gases, along with the chlorine content of the samples shown in Fig. 6. The samples shown in this figure are nonchloride plastics. PE, PS, or PET + NaCl meant a mixture of plastic sheet and NaCl, not a NaCl-impregnated plastic sheet. NaCl was used because it is the most abundant metal chloride in the environment.

Nonchloride plastics (PE, PS, and PET) produced dioxins even though their chlorine contents were negligible (<0.0005 from PE and PS sheet, 0.025 from PS foam, and 0.00 from PET). PET produced the greatest amount of dioxins (25.6 ng/g) among the nonchloride plastics combusted, followed by PE (11.8 ng/g). PS sheet gave only a trace amount (1.36 ng/g) of dioxins, whereas the PS form yielded an appreciable amount of dioxins (5.19 ng/g); this may be due to the amount of chlorine contamination. The foam form of plastics may absorb more contamination, and, consequently, produce more dioxins, than plastic sheet upon combustion. Addition of NaCl to PE reduced the total dioxin formation from 11.8 to 6.7 ng/g, mainly because of the reduction of PCDFs formation from 10.7 to 4.78 ng/g. On the other hand, a significant increase of total dioxin formation (from 1.36 to 17.7 ng/g) was observed in the case of PS. In the case of PET, only a slight increase of the total dioxin formation (from 25.6 to 28.9 ng/g) was recognized. However, PCDD formation increased considerably, from 0.566 to 7.242 ng/g.

When PE was burned with NaCl-NP, relatively high level of dioxins were formed (102 ng/g), compared with PE combusted with solid NaCl (6.07). An impregnated form of NaCl may release chlorine ions more readily than the solid form of NaCl. However, the results indicate that NaCl increases dioxin formation from plastics on combustion. It is reported that chloride sources in municipal wastes are 50% from NaCl and 45% from PVC (Kanters et al. 1996). Therefore, a role of PVC in dioxin formation from various materials was also investigated.

D. Polyvinyl Chloride Alone and with Other Plastics and Newspaper

PVC, with a chlorine content ranging from 35% to 55%, is one of the most widely used plastics. It is used in a wide range of consumer products, including packaging, cling film, bottles, credit cards, audio records, imitation leather, window frames, cables, pipes, wallpaper, and window blinds. PVC is difficult to recycle and consequently is combusted as a waste or ends up in landfills. Formation of dioxins from PVC on thermal degradation has been reported (Sinkkonen et al. 1995; Kanters et al. 1996; Wikström and Marklund 2000; Wagner and Green 1993). Also, cocombustion of PVC together with common plastics, including PE, PS, polyurethane, and epoxy-

and phenol/formaldehyde resin, produced dioxins in the high ng/g range (Wirts et al. 1998).

Figure 7 shows the results of dioxin analysis in the exhaust gases from the combustion of PVC and PVC with other plastics (PE, PS, and PET) and newspaper in an incinerator. The "×10" and "×100" indicate that the actual values are 10 and 100 times the values shown in the figure, respectively. The figure was constructed from the results of three articles (Yasuhara et al. 2001, 2005c; Katami et al. 2002). Table 4 shows the combustion chamber conditions and concentrations of O_2, CO, CO_2, and Cl ion in the exhaust gases, along with the chlorine content of the samples shown in Fig. 7. PVC was washed with hexane and ethanol to remove the phthalates used as a plasticizer. After removal of the phthalates, the chlorine content in the PVC was 51.3%. High levels of dioxin formation were expected from combustion of PVC because of its high chlorine content. The chamber and grate temperatures were adjusted to obtain low (42ppm) and high (880ppm) CO concentrations in the exhaust gases. The average chamber temperatures were 633°C for low CO concentration and 531°C for high CO concentration. The average grate temperatures were 900°C for low CO concentration and 742°C for high CO concentration. Generally, the lower the combustion temperature, the higher the CO concentration obtained. The results from the low CO concentration should be more comparable to the results from the rest of the samples than those from the high CO concentration because its combustion temperature is similar to those of others.

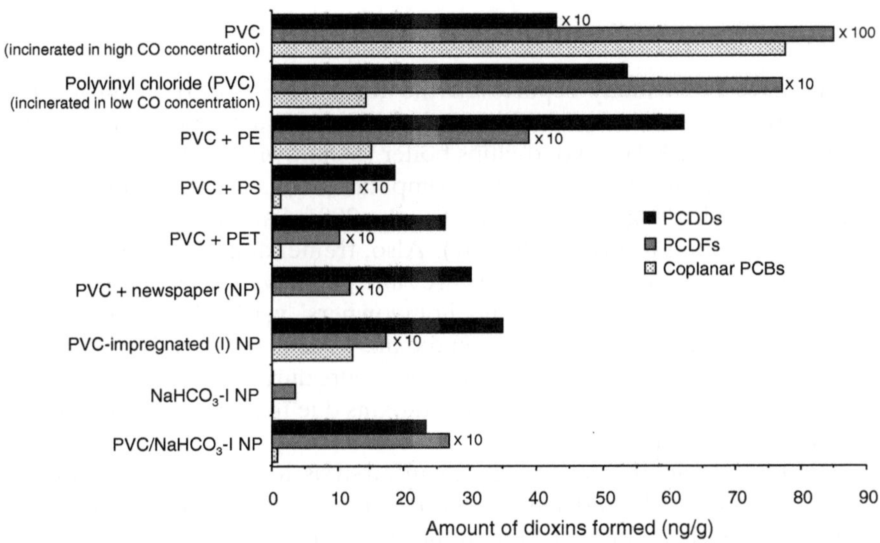

Fig. 7. Dioxin analysis of exhaust gases from combustion of PVC alone and PVC with other plastics and newspaper.

A mixed sheet of PVC with PE, PS, or PET (PVC + PE, PVC + PS, and PC + PET) was prepared from one PVC sheet sandwiched with two PE, PS, or PET sheets [PVC/PE, PS, or PET ≅ 1/10 (w/w)]. PVC-impregnated NP was prepared by soaking NP in a tetrahydrofuran solution of PVC and then allowing it to dry naturally at room temperature. $NaHCO_3$-impregnated NP was also soaked in a tetrahydrofuran solution of PVC and then dried naturally at room temperature to prepare PVC/$NaHCO_3$-impregnated NP.

Among a mixture of PVC and plastics or NP, PE produced the greatest amount of total dioxins (463 ng/g) when it was combusted with PVC, followed by PVC + PS (142 ng/g), PVC + PET (128 ng/g), and PVC + NP (103 ng/g). Even though the chlorine content of the PVC samples was approximately twice of those of the NaCl samples, these nonchloride plastic sheets produced dioxins much more with PVC (organic chloride source) than with NaCl (inorganic chloride source). It is interesting that NP produced more dioxins with impregnated PVC (220 ng/g) than with mixed PVC (103 ng/g). Impregnating $NaHCO_3$ apparently increased the dioxin formation from the PVC-I sample (288 ng/g). PVC itself is difficult to burn, but when it is combusted with other waste materials under poorly controlled conditions, such as low temperature and high CO concentration, dioxins will probably form in significant levels.

These results indicate that both organic and inorganic chloride were a source of chloride for dioxin formation. There are many reports on dioxin formation from the combustion of various waste materials with organic chloride, such as PVC (Christmann et al. 1989).

E. Wood and Leaves

There have been many reports on the formation of dioxins during the combustion of various woods in various systems: a fireplace in a home (Vikelsoe et al. 1994), a wood-chips boiler (Schatowitz et al. 1994; Samaras et al. 2001; Luthe et al. 1998; Pandompatam et al. 1997), an incinerator (Kolenda et al. 1994; Ikeguchi and Tanaka 2001; Salthammer et al. 1995), and an accidental fire (Carroll 2001). Also, tremendous amounts of fallen leaves have been burned after they are raked up in various areas, including public parks, streets, forests, and homeowners' yards. The smoke from burning leaves reportedly contained dioxins (Yasuhara et al. 2001). In particular, uncontrolled combustion, such as occurs during so-called "backyard burning" of domestic waste, produces dioxins due to inefficient combustion (Gullett et al. 2001).

Figure 8 shows the results of dioxin analysis in exhaust gases formed from various woods and leaves upon combustion. This figure was constructed from the results of four studies (Yasuhara et al. 2001, 2002, 2003; Katami et al. 2004a). The "×10" indicates that the actual values are 10 times the values shown in the figure. Table 5 shows the combustion chamber

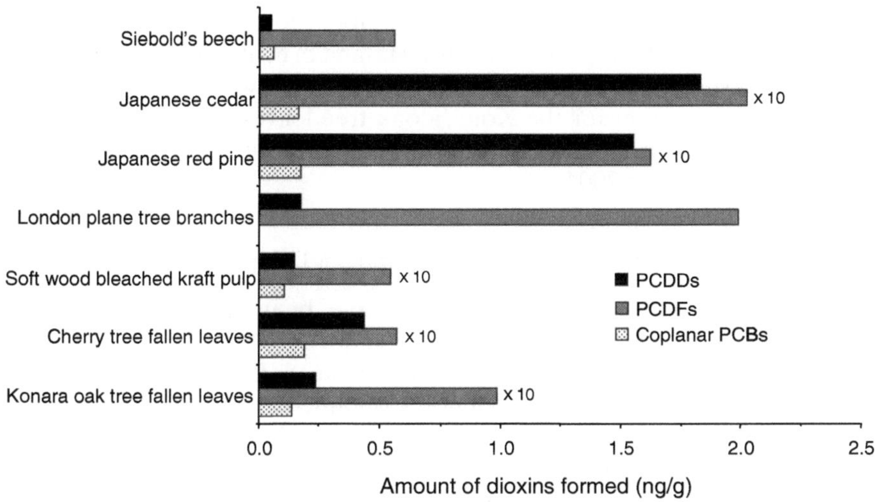

Fig. 8. Dioxin analysis of exhaust gases formed from combustion of wood and leaves.

conditions and concentrations of O_2, CO, CO_2, and Cl ion in the exhaust gases, along with the chlorine content of the samples shown in Fig. 8. These wood samples were prepared as follows: a log with bark (16 cm diameter) was cut into wood chips (5 cm^3) using an automatic circular saw. Among five different kinds of wood, a log of Japanese cedar produced the greatest amount of total dioxins (22.3 ng/g) upon combustion, followed by Japanese red pine (18.0 ng/g). The formation of dioxins is influenced by chlorine content as well as by combustion conditions (see Table 1). Various forms of wood, such as logs with/without bark, wood chips, and various kinds of bark and leaves, have been incinerated in both public and private facilities. Therefore, there are many reports on the formation of dioxins from various kinds of wood upon combustion. For example, wood chips and a mixture of peat (75%) and liquid packaging board (25%) produced 0.93 and 0.130 ng/g total dioxins, respectively, upon combustion at a temperature of at least 850°C (Sinkkonen et al. 1995). However, the exact composition of these wood samples, such as the chlorine content, is not consistent among the reports.

Fallen leaves were collected from the suburbs of a city and allowed to dry in the atmosphere for 2 mon. The leaves from a Konara oak tree and a cherry tree produced 10.2 and 6.37 ng/g dioxins upon combustion, respectively. The chlorine contents in the cherry tree leaves (0.074%) and in the Konara oak tree leaves (0.023%) did not correlate with total dioxin formation, suggesting that combustion temperature played a more important role in dioxin formation than the chlorine content. There is a report that chlorine content less than 1% in a combustion sample did not play a role in the dioxin formation (WikstrÖm et al. 1996; Hatanaka et al. 2000). The combustion chamber temperature for the cherry tree leaves was 746°C, which was much higher than that for the Konara oak tree leaves (692°C). Generally, it is reported that the higher the combustion temperature, the less the dioxin formation (Stanmore 2004).

F. Wood Impregnated with Seawater and Various Chlorides

As already mentioned, various woods have been combusted in large amounts both in private homes and in city incinerators. It is well known that the chloride content of combustion samples influences the formation of dioxins. More or less, wood samples obtained from various areas are contaminated with chlorides. Therefore, it is important to determine how dioxins form from various woods in the presence of chloride in high-temperature processes such as incineration to reduce their role in environmental contamination.

Figure 9 shows the results of dioxin analysis in exhaust gases formed from various kinds of wood impregnated with seawater, NaCl, chlordane, or pentachlorophenol upon combustion. This figure was constructed from the results of two studies (Yasuhara et al. 2002, 2003). The "×10" indicates that the actual values are 10 times the values shown in the figure. Table 6 shows the combustion chamber conditions and concentrations of O_2, CO, CO_2, and Cl ion in the exhaust gases, along with the chlorine content of the

Fig. 9. Dioxin analysis of exhaust gases formed from combustion of wood impregnated with seawater, NaCl, chlordane, or pentachlorophenol.

samples shown in Fig. 9. Bleached kraft pulp from soft wood with a chlorine content less than 0.01% was soaked in a NaCl solution for 10 min and then dried in an electric dryer at 110°C. The chlorine content of the NaCl-I bleached kraft pulp was 4.25%. Impregnating NaCl increased total dioxin formation from 0.799 ng/g (see Fig. 8) to 6.71 ng/g.

Seawater-impregnated pine and cedar were prepared as follows: a log with/without bark was placed in the sea at 20 m above the bottom for 3 mon. After the seawater-impregnated log was dried for 1 mon under standard conditions, the log was cut into small chips and then incinerated under the conditions shown in Table 1. Impregnating with seawater increased the total dioxin formation from 18.0 to 116.2 ng/g for Japanese red pine and from 22.3 to 98.58 ng/g for Japanese cedar (with bark). The results indicate that any wood used for construction that comes into contact with water—such as boats, ships, houses, and piers—produces significant amounts of dioxins on combustion and causes air contamination. For example, some logs collected near the sea contained a high level of chlorine and consequently produced relatively high levels of dioxins on combustion (Luthe et al. 1998; Pandompatam et al. 1997).

Waste woods were collected from beneath a floor where chlordane had been applied as an antitermite agent (chlordane content, 19.9 μg/g). The other waste woods were collected from beneath a floor where pentachlorophenol had been applied as a preservative (pentachlorophenol content, 0.42%). The waste woods contaminated with chlordane or pentachlorophenol produced 3.5 ng/g total dioxins upon incineration. The levels of total dioxins formed were rather low due to the high combustion temperatures (800°C; refer to Table 1) applied in this experiment, suggesting that dioxin formation can be controlled by raising the combustion temperature to more than 800°C. The role of combustion temperature in dioxin formation is discussed later.

G. Miscellaneous Materials

Many studies have been conducted on dioxin formation from various samples burned in small-scale incinerators, in addition to the results shown in Figures 4–9. Figure 10 shows the results of dioxin analysis in exhaust gases formed from various incinerated samples. This figure was constructed from the results of four studies (Yasuhara et al. 2005a,c, 2006; Katami et al. 2004b). The "×10," "100," and "×1,000" indicate that the actual values are 10, 100, 1,000 times the values shown in the figure, respectively. Table 7 shows the combustion chamber conditions and concentrations of O_2, CO, CO_2, and Cl ion in the exhaust gases, along with the chlorine content of the samples shown in Fig. 10.

One of the major sources of dioxins in the environment is the combustion of domestic waste materials. Domestic garbage is a mixture of tremendous numbers of substances, including various foods, household materials, and pet feces. For example, domestic waste, on a wet weight basis, from six

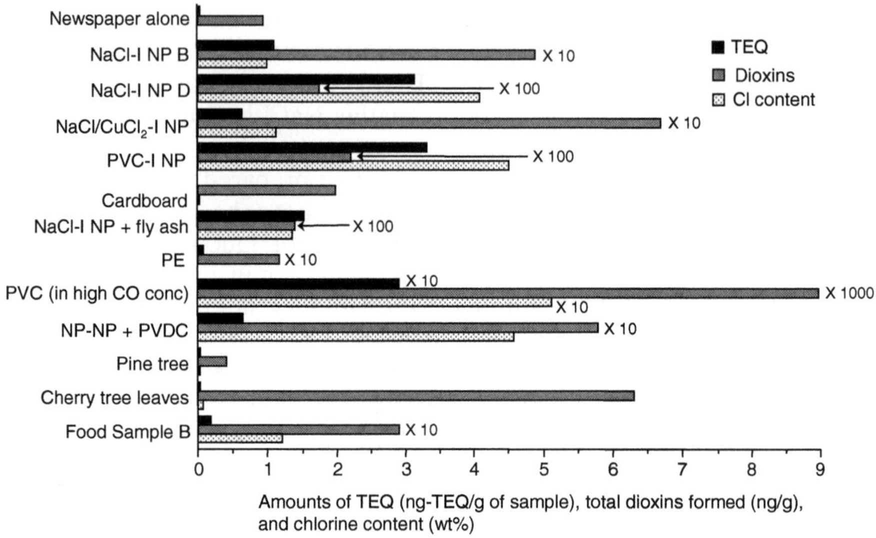

Fig. 10. Dioxin analysis of exhaust gases formed from combustion of various samples.

cities in Japan consisted of 38.3% food waste, 29.4% paper, 13.2% plastic, 3.9% glass, 3.5% metal, and 2.4% textiles in 1999 (JESC 2001). There has been great concern about the large amounts of garbage produced by domestic households in the modern world. This domestic waste has been combusted in large-scale incinerators under well-controlled conditions. Public concerns in developed countries, including the United States and Japan, however, have focused on the formation of toxic dioxins during the incineration of this garbage. Dioxin contamination in the environment occurs from the combustion of waste materials, as well as from many other high-temperature processes commonly used in industrial settings (Lustenhouwer et al. 1980; Tuppurainen et al. 1998). There are some reports on dioxin formation from domestic wastes incinerated under less-controlled systems (Gullett et al. 2001) and from waste plastic combusted at various temperatures (Yoneda et al. 2002).

Exhaust gases from an incinerator that contained mixtures of 67 food items—including fruits, vegetables, pasta, seafoods, meats, and processed foods and seasoned foods—were analyzed for dioxins (Katami et al. 2004b). Food sample A was a mixture of food scraps produced during food preparation and food sample B was a mixture of waste produced after eating. Food sample B produced more dioxins (29.1 ng/g) than did food sample A (18.9 ng/g), which may be due to the higher chlorine content in food sample B (1.22 wt%) than in food sample A (0.99 wt%). Also, $CaCl_2$ content was lower in food sample B (4.0 mg/L) than in food sample A (4.7 mg/L), whereas

NaCl content was higher in food sample B (5,900 mg/L) than in food sample A (5,100 mg/L). These results explain the higher formation levels of dioxins in food sample B than in food sample A.

Polyvinylidene chloride (PVDC), developed by Dow Chemical in the 1930s, was initially used as a protective coating to combat corrosion on military aircraft exposed to sea spray. Later, it was used as one of the components in a copolymer, and it soon found application as a film for wrapping food. PVDC is widely used as a thin barrier layer in multilayer sheets, films, and tubes. Eventually, PVDC is incinerated with other domestic waste materials. PVDC was combusted with a newspaper material with no printing (NP-NP) in a small-scale incinerator to examine dioxin formation in the exhaust gas. Addition of PVDC to NP-NP increased dioxin formation from 2.22 to 58.0 ng/g; chlorine content increased from 0.0064% to 4.59%. It has been reported that the level of dioxin formation from PVDC combusted at 750°C or lower was much higher than at 800°C or higher in a quartz tubular furnace (Ohta et al. 2001). When NP was combusted with PVDC, 1.525 ng/g dioxins was formed (Katsumata et al. 2003).

In the sample of electric wire coated with PVC, total PCDFs composed 85% of the total dioxins formed (38.3 ng/g) in the exhaust gas after combustion. The electric wire was composed of 37.6% copper wire and 62.4% wire coated with PVC. Therefore, the calculated value of the total dioxins without copper wire was 61.1 ng/g. In the case of PVC alone, PCDFs comprised 80% of the total dioxins formed (112 ng/g) in the exhaust gas. Based on these values, the presence of copper wire is seen to reduce dioxin formation by 30% from PCDDs, 58% from PCDFs, and 55% from the total dioxins (Yasuhara et al. 2005b). When PVC-cable sheathings were combusted, dioxins (ppm levels) were formed (Christmann et al. 1989).

Other miscellaneous samples also produced appreciable amounts of dioxins upon combustion: 4.10 ng/g from a PS plastic cover for an air conditioner, 1.98 ng/g from cardboard paper, and 8.10 ng/g from lightweight coated paper. These results suggest that waste materials containing plastic materials produce dioxins in the presence of chloride upon incineration.

VI. Toxicity Equivalence Quantity in Exhaust from Incinerated Samples

Dioxins are present in environmental and biological substances as extremely complex mixtures of various congeners. Some dioxins have been shown to cause toxic responses similar to those caused by 2,3,7,8-tetrachlorodibenzo-*p*-dioxin (TCDD), which is the most potent congener within these groups of compounds. Therefore, it is important to assess the risk caused by these chemicals for humans and wildlife. For this purpose, the concept of toxic equivalency factors (TEFs) has been developed and introduced to facilitate risk assessment and regulatory control of exposure to these mixtures. The method to calculate toxic equivalent (TEQ) concentrations in various environmental samples, including exhaust gases from incinerators, utilized TEF values (van den Berg et al. 1998).

Figure 11 shows toxicity equivalency quantity (ng-TEQ/g) of the representative samples shown in the tables. The values were obtained using the previously reported method (van den Berg et al. 1998). The "×10" indicates that the actual value is 10 times the value shown in the figure.

The samples that did not contain chlorine or were not burned with chlorides exhibited low levels of TEQ values, 0.013 from NP alone and 0.073 ng-TEQ/g from PE. In contrast, samples with high chlorine content, such as PVC (51.3%), gave a high level of TEQ values. Among the TEQ values from the samples reported in the 10 articles used as a core component of this review, PVC combusted in high CO content at low chamber temperature (531°C) gave the highest TEQ value of 29 ng-TEQ/g. On the other hand, when the same sample was combusted in low CO content at high temperature (633°C), the TEQ value was at the much lower level of 8.3 ng-TEQ/g. NP impregnated with either organic (PVC) or inorganic (NaCl) source resulted in similar values of TEQ (3.12 ng-TEQ/g from NaCl-I NP D and 3.31 ng-TEQ/g from PVC-I NP) and dioxin formation; this may be the result of the similar chlorine content in the combustion samples (4.08% in NaCl-I NP D and 4.51% in PVC-I NP). In the U.S., emissions of dioxins from PVC burned in house fires has been estimated at <1 g I-TEQ/yr (Carroll 2001).

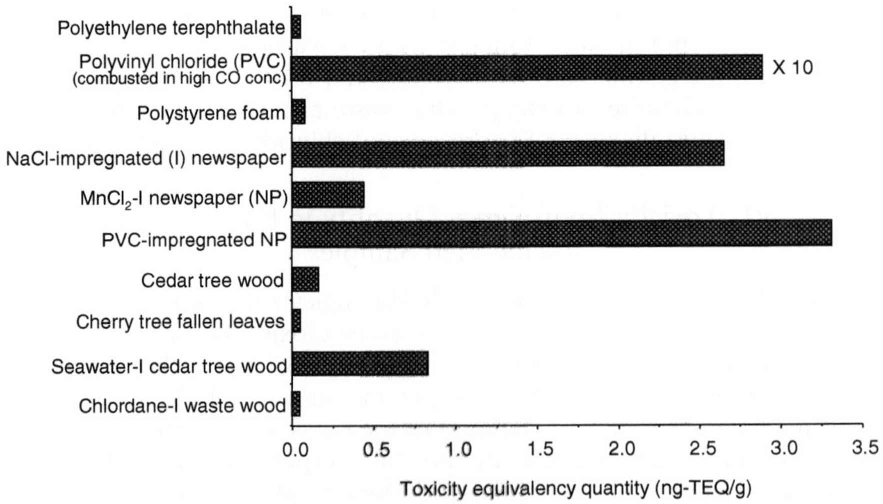

Fig. 11. Toxicity equivalency quantity (ng-TEQ/g) of representative samples from the samples shown in the tables.

The TEQ values in general correlated with the amount of chlorine content in combustion samples and the amount of dioxin formed in exhaust gases from an incinerator. When the same sample was, however, combusted at different temperatures, the sample combusted at low temperature (NaCl-I NP D) yielded a higher TEQ value (3.12 ng-TEQ/g) than did the sample combusted at high temperature (NaCl-I NP E; refer to Table 1) (0.026 ng-TEQ/g).

In the case of NaCl-I NP B (chlorine content, 0.99%), impregnating additional $CuCl_2$ (chlorine content of sample, 1.13%) reduced the TEQ value significantly from 1.08 to 0.62 ng-TEQ/g; nevertheless, it increased the dioxin formation in the exhaust gas from 49 to 67 ng/g.

The TEQ value of the pine wood sample was calculated in ng/m^3 to compare the results with the other data. When pine was combusted at below 600°C and at over 860°C, The TEQ values of its exhaust gas were 2.9 $ng\text{-}TEQ/m^3$ and 0.0023 $ng\text{-}TEQ/m^3$, respectively. Similar results were reported from waste wood combusted at 623°C (0.86 $ng\text{-}TEQ/m^3$) and at 1,086°C (0.14 $ng\text{-}TEQ/m^3$) (Ikeguchi and Tanaka 2001). NaCl-I bark (chlorine content, 0.76%) gave 3.2 $ng\text{-}TEQ/m^3$ upon combustion (Pandompatam et al. 1997). Another report indicated that waste woods gave TEQ values of 2.7–14.4 $ng\text{-}TEQ/m^3$ (Kolenda et al. 1994). Chlordane- and pentachlorophenol-I waste woods gave 0.34 and 0.47 $ng\text{-}TEQ/m^3$, respectively (Yasuhara et al. 2003). These results indicate that wood samples may produce certain amounts of toxic materials with significant TEQ values upon combustion. In fact, seven actual wood-burning facilities were found to yield 0.004–9.820 ng I-TEQ/Nm^3, with the majority of values >0.1 ng ITEQ/Nm^3. The highest concentrations occurred after the addition of halogenated materials (NH_4Cl-hardened or PVC-coated plywood) to the input of the incinerator in this study (Kolenda et al. 1994).

Samples of fallen leaves, which produced 6–10 ng/g levels of dioxins, gave TEQ values of 0.12–0.48 $ng\text{-}TEQ/m^3$ (Katami et al. 2004a). These values are slightly higher than the maximum allowable level (0.1 $ng\text{-}TEQ/m^3$) in the exhaust gases set by the government of Japan. Therefore, it is important to incinerate fallen leaves in well-controlled systems to minimize the formation of dioxins. Furthermore, burning fallen leaves in a homeowner's yard or in a public park is not recommended.

VII. Dioxin Formation Mechanisms During Incineration

There have been many studies on the formation of dioxins under various conditions (Kanters et al. 1996; Wikström and Marklund 2000; Iino et al. 1999, 2000). However, formation mechanisms of dioxins from waste materials upon combustion in an incinerator are not yet completely understood because many complex reaction pathways seem to be involved in the process (Huang and Buekens 1995). Some of the possible reaction mechanisms of dioxin formation upon combustion of waste materials have been proposed in the past two decades, including high-temperature pyrosynthesis, low-temperature de novo formation from macromolecular carbon and organic or inorganic chlorine present in the fly ash matrix, and formation from different organic precursors such as chlorophenols (Tuppurainen et al. 1998). It is reported that dioxins form from precursors via organic chemical reactions, such as the condensation reaction of two molecules of chlorophenols and the cyclization reaction of polychlorinated biphenyls (Luijk et al. 1994). It is also hypothesized that dioxin formation involves a radical reaction between simple carbon radicals and chloride radicals under high-temperature conditions (Stieglitz and Vogg 1987; Huang and Buekens 1995). A study on the effect of temperature on PVC combustion in a downstream tubular furnace resulted in a close correlation ($R^2 = 0.97$) between temperature and dioxins formation (Kim et al. 2004).

The effect of combustion temperatures on dioxin formation was reportedly different between the combustion chambers. Also, formation of dioxin varied in different sections of municipal waste incinerators. The amounts of dioxins were significantly reduced when HCl was not supplied to the main combustion section, suggesting that dioxins were formed mainly in the main combustion section (Hatanaka et al. 2005). Several statistically significant relationships between HCl emissions and the emissions of dioxins were also reported (Wagner and Green 1993). On the other hand, the addition of HCl had little or not apparent effect on the level of dioxin formation during PVDC combustion (Ohta et al. 2004). When wood chips were combusted with PVC resin or NaCl, chloride content did not influence dioxin formation at 1,000°–1,100°C but it increased it significantly at lower temperatures (Ikeguchi and Tanaka 2001).

There are several excellent reviews on dioxin formation including the role of temperature (Lustenhouwer et al. 1980; Stanmore 2004). There are generally two reactions associated with dioxin formation from waste materials based on different combustion temperatures. One reaction is a homogeneous reaction that occurs at 500°–800°C. Under these temperatures, dioxin formation and degradation may occur simultaneously in the gas phase. Possible reaction mechanisms occurring during the homogeneous reaction have been hypothesized according to general organic chemistry, including the cyclization of polychlorobiphenyls and polychlorodiphenyl ethers, the chlorination of dibenzofuran, and the dechlorination of octachloro dibenzofuran (Choudry and Hutzinger 1983; Ballschmiter et al. 1985). Most studies on dioxin formations from incinerators, including the ones discussed in Figures 4–10, were conducted at the temperatures that promote a homogeneous reaction. The other reaction is a heterogeneous reaction that occurs at 200°–400°C. Dioxins are formed on solid surfaces such as fly ash and soot by some catalysis (metals) during this reaction. It has been proposed that the precursors of dioxins in this reaction are aromatic chlorides, such as chlorophenols and chlorobenzenes, and elemental carbons *via de novo* reaction (Addink and Olie 1995). Dioxin formation in municipal waste incineration systems was, therefore, hypothesized as a two-stage process: (1) the formation of the graphitic structure of soot particles in the combustion zone (heterogeneous reaction); and (2) the conversion of the graphitic structure of soot particles to aromatic compounds, including dioxins in the postcombustion zone at high temperatures (homogeneous reaction) (Huang and Buekens 1995).

From the experiment conducted using a laboratory-scale fluidized-bed reactor, of which each temperature of primary and secondary combustion zones was set independently to 700°, 800°, and 900°C, it was found that dioxin concentration was reduced as the temperature of the secondary combustion zone increased. On the other hand, as the temperature of the primary combustion zone increased, the dioxins concentration also increased, suggesting that keeping the temperature of the secondary combustion zone high and primary combustion zone low reduces the release of dioxins during waste incineration (Hatanaka et al. 2001a). There are many reports suggesting that the chlorine content of combustion samples and the combustion temperature play an important role in dioxin formation in municipal incinerators (Kawabata et al. 2002, 2003; Hatanaka et al. 2001b; Procaccini et al. 2003).

Figure 12 shows plots of dioxin formation against chamber temperatures. The data used are those reported in the references shown in Tables 1–7. No significant relationship between dioxin formation and chamber temperatures was observed. However, it is obvious that dioxin formation occurred

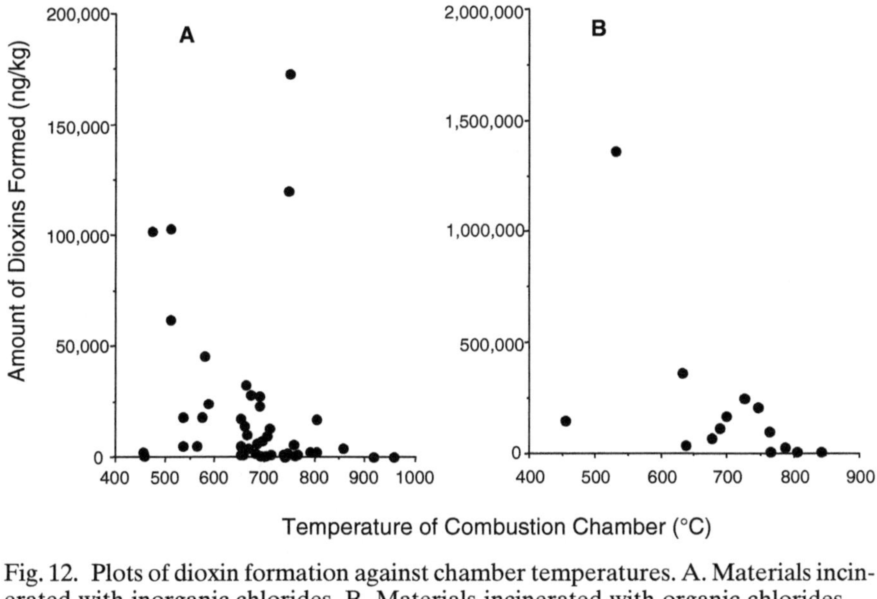

Fig. 12. Plots of dioxin formation against chamber temperatures. A. Materials incinerated with inorganic chlorides. B. Materials incinerated with organic chlorides.

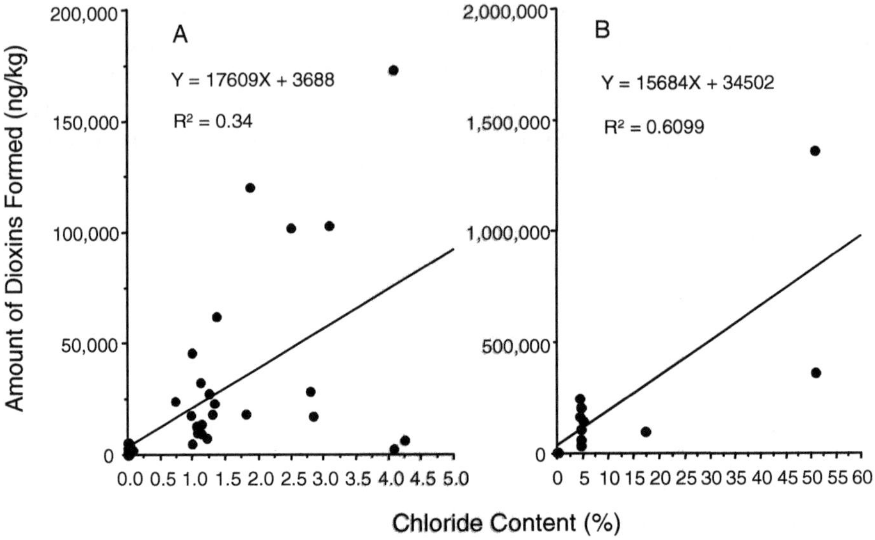

Fig. 13. Relationship between chloride content of incinerated samples and dioxin formation. A. Materials incinerated with inorganic chlorides. B. Materials incinerated with organic chlorides.

at temperatures above 450°C and reduced significantly at temperatures above 850°C.

Combustion temperatures may play an important role in dioxin formation in exhaust gases from incineration of waste materials. However, the reaction occurring in an incinerator is extremely complex and there are many factors in addition to combustion temperature influencing dioxin formation. Factors involved in dioxin formation are chorine content, number of benzene rings, the level of metals in samples, and O_2, CO, and HCl concentration in the combustion chamber. As already mentioned, the chloride content (%) in a combustion sample plays some role in dioxin formation (ng/kg). Figure 13 shows the relationship between chloride content in the combustion samples and dioxin formation. The data used are those reported in the references shown in Tables 1–7. This figure suggests that there is no significant relationship between dioxin formation and chloride content in the data obtained from the references shown in Tables 1–7.

Summary

There has been great concern about dioxins—polychlorinated dibenzo dioxins (PCDDs), polychlorinated dibenzo furans (PCDFs), and polychlorinated biphenyls (PCBs)—causing contamination in the environment because the adverse effects of these chemicals on human health have been known for many years. Possible dioxin-contamination has received much attention recently not only by environmental scientists but also by the public, because dioxins are known to be formed during the combustion of industrial and domestic wastes and to escape into the environment *via* exhaust gases from incinerators. Consequently, there is a pressing need to investigate the formation mechanisms or reaction pathways of these chlorinated chemicals to be able to devise ways to reduce their environmental contamination.

A well-controlled small-scale incinerator was used for the experiments in the core references of this review. These articles report the investigation of dioxin formation from the combustion of various waste-simulated samples, including different kinds of paper, various kinds of wood, fallen leaves, food samples, polyethylene (PE), polystyrene (PS), polyvinyl chloride (PVC), polyvinylidene chloride, polyethylene tetraphthalate (PET), and various kinds of plastic products. These samples were also incinerated with inorganic chlorides (NaCl, KCl, $CuCl_2$, $MgCl_2$, $MnCl_2$, $FeCl_2$, $CoCl_2$, fly ash, and seawater) or organic chlorides (PVC, chlordane, and pentachlorophenol) to investigate the role of chlorine content and/or the presence of different metals in dioxin formation. Some samples, such as newspapers, were burned after they were impregnated with NaCl or PVC, as well as being cocombusted with chlorides. The roles of incineration conditions, including chamber temperatures, O_2 concentrations, and CO concentrations, in dioxin formation were also investigated. Dioxins (PCDDs, PCDFs, and coplanar-PCBs) formed in the exhaust gases from a controlled small-scale incinerator, where experimental waste samples were burned, were analyzed by gas chromatography/mass spectrometry.

Formation of total PCDFs was much higher than that of PCDDs in all samples. The total PCDFs comprised 70%–90% of the total dioxin formed. The amount of total PCDFs formed ranged from 0.78 ng/g (newspaper) to 8,490 ng/g (PVC burned in high CO concentration). The amount of total PCDDs formed ranged from 0.02 ng/g (newspaper) to 430 ng/g (PVC). Coplanar PCBs were found at the lowest level of the dioxins formed. Their formation levels ranged from 0 ng/g (newspaper) to 77.6 ng/g (PVC). It is obvious that the samples with either inorganic or organic chlorides produced much more dioxins than the sample without chlorides when incinerated under similar conditions. It is not clear how inorganic and organic chloride contribute differently to dioxin formation. Among the metals examined, copper seems to have higher activity toward dioxin formation than other metals. It acted not only as a catalyst but also as a transmitter of heterogeneous chlorine.

The toxicity equivalence quantity (TEQ) values generally correlated with the amount of chlorine content in the samples and the amount of dioxin formed in exhaust gases from an incinerator. When the same sample was incinerated at different temperatures, however, the sample burned at low temperature yielded a higher TEQ value than did the sample burned at high temperature. The samples that did not contain chlorine or were not combusted with chlorides exhibited low TEQ values. In contrast, samples with high chlorine content, such as PVC (51.3%), gave high TEQ values.

Combustion temperatures may play an important role in dioxin formation in exhaust gases from the incineration of waste materials. However, no significant relationship between dioxin formation and chamber temperatures was reported in the core articles. However, it is obvious that dioxin formation occurred at temperatures above 450°C and was reduced significantly at temperatures above 850°C. The reaction occurring in an incinerator is extremely complex, and there are many factors in addition to combustion temperature influencing dioxin formation. Even though it is possible to hypothesize reasonable formation mechanisms of dioxins produced in exhaust gases according to the results obtained from experiments in classical chemistry, the reactions involved in an incinerator are extremely complex and heterogeneous. More detailed investigation of the many individual factors influencing dioxin formation is needed to find ways to reduce their formation in individual and municipal incinerators.

References

Addink R, Olie AJ (1995) Mechanisms of formation and destruction of polychlorinated dibenzo-*p*-dioxins and dibenzofurans in heterogeneous systems. Environ Sci Technol 29:1425–1435.

Ballschmiter K, Zoller W, Buchert H (1985) Th. Class Fresenius Z. Anal Chem 322:587–594.

Bingham AG, Edmunds CJ, Graham BWL, Jones MT (1989) Determination of PCDDs and PCDFs in car exhaust. Chemosphere 19:669–673.

Brown RP, Cooper KR, Cristini A, Rappe C, Bergqvist P-A (1994) Polychlorinated dibenzo-*p*-dioxins and dibenzofurans in *Mya arenaria* in the Newark/Raritan Bay Estuary. Environ Toxicol Chem 13:523–528.

Carroll WF, Jr (2001) The relative contribution of wood and poly (vinyl chloride) to emissions of PCDD and PCDF from house fires. Chemosphere 45: 1173–1180.

Choudry GG, Hutzinger O (1983) Mechanistic Aspects of the Thermal Formation of Halogenated Organic Compounds Including Polychlorinated Dibenzodioxins. Gordon and Breach, New York.

Christmann W, Kasiske D, Kloeppel KD, Partscht H, Rotard W (1989) Combustion of polyvinylchloride. An important source for the formation of PCDD/PCDF. Chemosphere 19:387–392.

Coulston F, Pocchiari F (1983) Accidental Exposure to Dioxins: Human Health Aspects. Academic Press, New York.

Fängmark I, Strömberg B, Berge N, Rappe C (1994) Influence of small fly ash particles on the post-combustion formation of PCDDs, PCDFs, PCBs and CPs in a pilot incinerator. Chemosphere 29:1903–1909.

Fukunaga I (1998) Technical trend of municipal solid waste treatment in Europe. Waste Manag Res 9:487–495.

Giugliano M, Cernuschi S, Ghezzi U (1989) The emission of dioxins and related compounds from the incineration of municipal solid wastes with high contents of organic chlorine (PVC). Chemosphere 19:407–411.

Gullett BK, Bruce KR, Beach LO (1990) The effect of metal catalysts on the formation of polychlorinated dibenzo-*p*-dioxin and polychlorinated dibenzofuran precursors. Chemosphere 20:1945–1952.

Gullett BK, Bruce KR, Beach LO, Drago AM (1992) Mechanistic steps in the production of PCDD and PCDF during waste combustion. Chemosphere 25:1387–1392.

Gullett BK, Lemieux PM, Lutes CC, Winterrowd CK, Winters DL (2001) Emissions of PCDD/F from uncontrolled, domestic waste burning. Chemosphere 43:721–725.

Hatanaka T, Imagawa T, Takeuchi M (2000) Formation of PCDD/Fs in artificial solid waste incineration in a laboratory-scale fluidized-bed reactor: influence of contents and forms of chlorine sources in high-temperature combustion. Environ Sci Technol 34:3920–3924.

Hatanaka T, Imagawa T, Kitajima A, Takeuch M (2001a) Effects of combustion temperature on PCDD/Fs formation in laboratory-scale fluidized-bed incineration. Environ Sci Technol 35:4936–4940.

Hatanaka T, Imagawa T, Kitajima A (2001b) PCDD/Fs formation in laboratory-scale fluidized-bed incineration: influence of chlorine content in low-temperature combustion. Organohalogen Compd 50:430–433.

Hatanaka T, Kitajima A, Takeuchi M (2005) Role of chlorine in combustion field in formation of polychlorinated dibenzo-*p*-dioxins and dibenzofurans during waste incineration. Environ Sci Technol 39:9452–9456.

Huang H, Buekens A (1995) On the mechanisms of dioxin formation in combustion processes. Chemosphere 31:4099–4117.

Iino F, Imagawa T, Takeuchi M, Sadakata M (1999) De novo synthesis mechanism of polychlorinated dibenzofurans from polycyclic aromatic hydrocarbons and the characteristic isomers of polychlorinated naphthalenes. Environ Sci Technol 33:1038–1043.

Iino F, Imagawa T, Gullett BK (2000) Dechlorination-controlled polychlorinated dibenzofuran isomer patterns from municipal waste incinerators. Environ Sci Technol 34:3143–3147.

Ikeguchi T, Tanaka M (2001) Dioxin emissions from industrial wood waste incinerators. Organohalogen Compd 50:390–393.

Inui T, Otowa T, Tsutchihashi K, Takegami Y (1982) Complete oxidation of active carbon at low temperatures by composite catalysts. Carbon 20:213–217.

Jay K, Stieglitz L (1991) On the mechanism of formation of polychlorinated aromatic compounds with copper(II) chloride. Chemosphere 22:987–996.

JESC (2001) Fact Book: Waste Management & Recycling in Japan. Japan Environmental Sanitation Center, Tokyo, Japan, p 22.

JMHW (1997) Standard method for determination of dioxins in waste management. Notification no. 234, Japan Ministry of Health and Welfare, Tokyo, Japan.

Kanters MJ, van Nispen R, Louw R, Mulder P (1996) Chlorine input and chlorophenol emission in the lab-scale combustion of municipal solid waste. Environ Sci Technol 30:2121–2126.

Katami T, Ohno N, Yasuhara A, Shibamoto T (2000) Formation of dioxins from sodium chloride-impregnated newspapers by combustion. Bull Environ Contam Toxicol 64:372–376.

Katami T, Yasuhara A, Okuda T, Shibamoto T (2002) Formation of PCDDs, PCDFs, and coplanar PCBs from polyvinyl chloride during combustion in an incinerator. Environ Sci Technol 36:1320–1324.

Katami T, Yasuhara A, Shibamoto T (2004a) Formation of dioxins from incineration of fallen leaf. Bull Environ Contam Toxicol 72:114–118.
Katami T, Yasuhara A, Shibamoto T (2004b) Formation of dioxins from incineration of foods found in domestic garbage. Environ Sci Technol 38:1062–1065.
Katsumata H, Yabuuchi M, Kaneco S, Suzuki T, Ohta K (2003) Formation of dioxins from newspaper combustion in the presence of sodium chloride and polyvinylidene chloride. Fresenius Environ Bull 12:406–412.
Kawabata H, Usui T, Marukawa K, Hara S, Nakazato H, Tanaka T (2002) Effect of chlorine sources and forms on formation of dioxins/furans in the combustion process. J Jpn Soc Waste Manag Expert 13:184–192.
Kawabata H, Usui T, Marukawa K, Hara S, Nakazato H, Tanaka T, Ono-Nakazato H (2003) Mechanism of dioxins/furans formation at high temperature in combustion processes. ISIJ Int 43:461–467.
Kim K-S, Hong K-H, Ko Y-H, Kim M-G (2004) Emission characteristics of PCDD/Fs, PCBs, chlorobenzenes, chlorophenols, and PAHs from polyvinylchloride combustion at various temperatures. J Air Waste Manag Assoc 54:555–562.
Kleopfer RD, Greenall RL, Viswanathan TS, Kirchmer CJ, Gier A, Muse J (1989) Determination of polychlorinated dibenzo-*p*-dioxins and dibenzofurans in environmental samples using high resolution mass spectrometry. Chemosphere 18:109–118.
Kolenda J, Gass H, Wilken M, Jager J, Zeschmar-Lahl B (1994) Determination and reduction of PCDD/F emissions from wood burning facilities. Chemosphere 29:1927–1938.
Kuzuhara S, Sato H, Kasai E, Nakamura T (2003) Influence of metallic chlorides on the formation of PCDD/Fx during low-temperature oxidation of carbon. Environ Sci Technol 37:2431–2435.
Luijk R, Akkerman DM, Slot P, Olle K, Kapteljn F (1994) Mechanism of formation of polychlorinated dibenzo-*p*-dioxins and dibenzofurans in the catalyzed combustion of carbon. Environ Sci Technol 28:312–321.
Lustenhouwer JWA, Olite K, Hutzinger O (1980) Chlorinated dibenzo-*p*-dioxins and related compounds in incinerator effluents. A review of measurements and mechanisms of formation. Chemosphere 9:501–522.
Luthe C, Karidio I, Uloth V (1998) Dioxins formation in salt-laden power boilers: a mass balance. Chemosphere 36:231–249.
Marquis PJ, Hackett M, Holland LG, Larsen ML, Butterworth B, Kuehl DW (1994) Analytical methods for a national study of chemical residues in fish. Chemosphere 29:495–508.
Nagata K (1992) State-of-the-art of PCDD/PCDF emission control for municipal waste incinerators. Waste Manag Res 3:217–237.
Oehme M, Kirschmer P (1984) Isomer-selective determination of tetrachlorodibenzo-*p*-dioxins using hydroxyl negative ion chemical ionization mass spectrometry combined with high-resolution gas chromatography. Anal Chem 56:2764–2769.
Ohta M, Oshima S, Iwasa T, Osawa N, Kumatoriya K, Yamazaki A, Takasuga T, Matsushita M, Umedzu N (2001) Formation of PCDDs and PCDFs during the combustion of polyvinylidene chloride. Chemosphere 44:1389–1394.

Ohta M, Oshima S, Osawa N, Iwasa T, Nakamura T (2004) Formation of PCDDs and PCDFs during the combustion of polyvinylidene chloride and other polymers in the presence of HCl. Chemosphere 54:1521–1531.

Pandompatam B, Kumar Y, Guo I, Linen AJ (1997) Comparison of PCDD and PCDF emissions from hog fuel boilers and hospital waste incinerators. Chemosphere 34:1065–1073.

Procaccini C, Bozzelli JW, Longwell JP, Sarofim AF, Smith KA (2003) Formation of chlorinated aromatics by reactions of Cl·, Cl_2, and HCL with benzene in the cool-down zone of a combustor. Environ Sci Technol 37:1684–1689.

Rappe C, Buser HR (1980) Chemical properties and analytical methods. In: Kimbrough RD (ed) Halogenated Biphenyls, Terphenyls, Naphthalenes, Dibenzo-*p*-Dioxins, and Related Products. Elsevier/North Holland Biomedical Press, New York, pp 48–68.

Safe S, Bandiera S, Sawyer T, Robertson L, Sate L, Parkinson A, Thomas PE, Ryan DE, Reik LM, Levin W, Denomme MA, Fujita T (1985) PCBs: structure–function relationships and mechanism of action. Environ Health Perspect 60:47–56.

Salthammer T, Klipp H, Peek R, Marutzky R (1995) Formation of polychlorinated dibenzo-*p*-dioxins (PCDD) and polychlorinated dibenzofurans (PCDF) during the combustion of impregnated wood. Chemosphere 30:2051–2060.

Samaras P, Skodras G, Sakellaropoulos GP, Blumenstock M, Schramm K-W, Kettrup A (2001) Toxic emissions during co-combustion of biomass–waste wood–lignite blends in an industrial boiler. Chemosphere 43:751–755.

Schatowitz B, Brandt G, Gafner F, Schlumpe E, Buhler R, Hasler P, Nussbaumer T (1994) Dioxin emissions from wood combustion. Chemosphere 29: 2005–2013.

Sinkkonen S, Mäkelä R, Vesterinen R, Lahtiperä M (1995) Chlorinated dioxins and dibenzothiphenes in fly ash samples from combustion of peat, wood chips, refuse derived fuel and liquid packaging boards. Chemosphere 31:2629–2635.

Stanmore BR (2004) The formation of dioxins in combustion systems. Combust Flame 136:398–427.

Stieglitz L, Vogg H (1987) On formation conditions of PCDD/PCDF in fly ash from municipal waste incinerators. Chemosphere 16:1917–1922.

Stieglitz L, Jay K, Hell K, Wilhelm J, Polzer J, Buekens A (2003) Investigation of the formation of polychlorodibenzodioxins *l*-furans and of other organochlorine compounds in thermal indusrial processes. Wissenschaf Be-Forsch Karlsr i–v:1–48.

Takasugi T, Makino T, Tsubota K, Takeda N (2000) Formation of dioxins (PCDDs/ PCDFs) by dioxin-free fly ash as a catalyst and relation with several chlorine-sources. Chemosphere 40:1003–1007.

Tiernan TO, Garrett JH, Solch JG, Wagel DJ, Van Ness GF, Taylor NL (1989) Improved separations procedures for isolating 2,3,7,8-TCDD and 2,3,7,8-TCDF from chemically-complex aqueous and solid sample matrices and for definitive quantitation of these isomers at ppq to ppt concentrations. Chemosphere 18:93–100.

Tong HY, Monson SJ, Gross ML, Bopp RF, Simpson HJ, Deck BL, Moser FC (1990) Analysis of dated sediment samples from the Newark Bay area for selected PCDD/Fs. Chemosphere 20:1497–1502.

Tuppurainen K, Halonen I, Ruokojärvi P, Tarhanen J, Ruuskanen J (1998) Formation of PCDDs and PCDFs in municipal waste incineration and its inhibition mechanisms: a review. Chemosphere 36:1493–1511.

van den Berg M, Birnbaum L, Bosveld ATC, Brunström B, Cook P, Feeley M, Giesy JP, Hanberg A, Hasegawa R, Kennedy SW, Kubiak T, Larsen J, Rolaf van Leeuwen FX, Djien Liem AK, Nolt C, Peterson RE, Poellinger L, Safe S, Schrenk D, Tillitt D, Tysklind M, Younes M, Wærn F, Zacharewski T (1998) Toxic equivalency factors (TEFs) for PCBs, PCDDs, PCDFs for humans and wildlife. Environ Health Perspect 106:775–792.

Vikelsoe J, Madsen H, Hansen K (1994) Emission of dioxins from Danish woodstoves. Chemosphere 29:2019–2027.

Wagner JC, Green AES (1993) Correlation of chlorinated organic compound emissions from incineration with chlorinated organic input. Chemosphere 26:2039–2054.

Wikström E, Marklund S (2000) Secondary formation of chlorinated dibenzo-*p*-dioxins, dibenzofurans, biphenyls, benzenes, and phenols during MSW combustion. Environ Sci Technol 34:604–609.

Wikström E, Löfvenius G, Rappe C, Marklund S (1996) Influence of level and form of chlorine on the formation of chlorinated dioxins, dibenzofurans, and benzenes during combustion of an artificial fuel in a laboratory reactor. Environ Sci Technol 30:1637–1644.

Williams LL, Giesy JP (1995) Polychlorinated biphenyls and 2,3,7,8-tetrachlorodibenzo-*p*-dioxin equivalents in eggs of double-crested cormorants from a colony near Green Bay, Wisconsin, USA. Arch Environ Contam Toxicol 29:327–333.

Wirts M, Lorenz W, Bahadir M (1998) Does co-combustion of PVC and other plastics lead to enhanced formation of PCDD/F? Chemosphere 37:1489–1500.

Yasuhara A, Katami T, Okuda T, Ohno N, Shibamoto T (2001) Formation of dioxins during the combustion of newspapers in the presence of sodium chloride and poly(vinyl chloride). Environ Sci Technol 35:1373–1378.

Yasuhara A, Katami T, Okuda T, Shibamoto T (2002) Role of inorganic chloride in formation of PCDDs, PCDFs, and coplanar PCBs from combustion of plastics, newspaper, and pulp in an incinerator. Environ Sci Technol 36: 3924–3927.

Yasuhara A, Katami T, Shibamoto T (2003) Formation of PCDDs, PCDFs, and coplanar PCBs from incineration of various woods in the presence of chlorides. Environ Sci Technol 37:1563–1567.

Yasuhara A, Katami T, Shibamoto T (2004) Formation of dioxins from incineration of foods found in domestic garbage. Environ Sci Technol 38:1062–1065.

Yasuhara A, Katami T, Shibamoto T (2005a) Dioxin formation during combustion of nonchloride plastic, polystyrene and its product. Bull Environ Contam Toxicol 74:899–903.

Yasuhara A, Tanaka Y, Katami T, Shibamoto T (2005b) The role of metals in dioxin formation from combustion of newspapers and polyvinyl chloride in an incinerator. Chemosphere 58:891–896.

Yasuhara A, Katami T, Shibamoto T (2005c) A role of alkaline elements in formation of PCDDs, PCDFs, and coplanar PCBs during combustion of various paper samples. J Environ Sci Health 40:991–1001.

Yasuhara A, Katami T, Shibamoto T (2006) Formation of dioxins from combustion of polyvinylidene chloride in a well-controlled incinerator. Chemosphere 62: 1899–1906.

Yoneda K, Ikeguchi T, Yagi Y, Tamade Y, Omori K (2002) A research on dioxin generation from the industrial waste incineration. Chemosphere 46:1309–1319.

Manuscript received April 13; accepted April 20, 2006.

Rev Environ Contam Toxicol 190:43–125

Coca and Poppy Eradication in Colombia: Environmental and Human Health Assessment of Aerially Applied Glyphosate

Keith R. Solomon, Arturo Anadón, Gabriel Carrasquilla, Antonio L. Cerdeira, Jon Marshall, and Luz-Helena Sanin

Contents

Communicated by G.W. Ware.

K.R. Solomon (✉)
Centre for Toxicology and Department of Environmental Biology, University of Guelph, Guelph, ON, N1G 2W1, Canada

A.L. Cerdeira
EMBRAPA, Ministry of Agriculture, Jaguariuna, SP 13820-000, Brazil

A. Anadón
Departamento de Toxicología y Farmacología, Facultad de Veterinaria, Universidad Complutense de Madrid, 28040 Madrid, Spain

G. Carrasquilla
Hospital Universitario, Fundación Santa Fe de Bogotá, Calle 116 No. 9, Bogotá

E.J.P. Marshall
Marshall Agroecology Limited, 2 Nut Tree Cottages, Barton, Winscombe, Somerset, BS25 1DU, UK

L.-H. Sanin
Department of Public Health Sciences, Faculty of Medicine, University of Toronto, Toronto, ON, M5S 1A8, Canada *and*
Autonomous University of Chihuahua
National Institute of Public Health, Mexico

I. Introduction

It is estimated that some 200 million people worldwide use illicit drugs. Most of these drugs have natural origins, such as cannabis, cocaine, and the opiates; however, the synthetic drugs such as the amphetamines also comprise a significant proportion of these uses (UNODC 2003). In response to the socioeconomic impacts of the production and distribution of illicit drugs, a number of individual nations, as well as multinational organizations, have initiated programs to reduce and eventually eliminate their production and distribution (UNODC 2003).

Coca (*Erythroxylum coca* and related species) is commonly associated with the tropical regions of South America. A number of species of coca are found in South America, and various varieties grow in the wild or are cultivated in different climatic conditions. It is primarily found in regions with temperatures above 25°C and with rainfall >1000 mm/yr. Currently, it is widely cultivated in Colombia, Bolivia, and Peru, with some cultivation in Ecuador, Venezuela, Brazil, and Argentina. In 2002, it was estimated that, of the 800 t cocaine produced in Latin America, 580 t was produced in Colombia, a reduction of about 100 t from 1999 (UNODC 2003). It is estimated that, worldwide, about 14 million people abuse cocaine (UNODC 2003).

Opium, morphine, and its derivative, heroin, are produced from the poppy *Papaver somniferum,* which is primarily grown in Asia. Global production of opium in 2002 was estimated to be 4,500 t, of which about 97 t was produced in South America; of this, 50 t was produced in Colombia (UNODC 2003). It is estimated that, globally, about 15 million people use opiates and that about 10 million of these use heroin (UNODC 2003). Similar to coca, the use of opium and morphine has historical roots in the traditional society of the producer regions but became more widely used as a human medicine when introduced to other parts of the world. Although morphine is still used for medicinal purposes, heroin use is largely illegal and its production and distribution have significant socioeconomic impacts in producer and consumer nations.

The growing and production of illicit drugs in Colombia have significant political, social, economic, and environmental impacts. While recognizing the importance of the political, social, and economic aspects of the issue, this review is focused on the human health and environmental significance of growing these crops and the control of coca and poppy through the use of the aerially applied herbicide glyphosate.

It is important to recognize that the actual production of coca and poppy as well as the processing and production of cocaine and heroin involve significant environmental impacts. Both coca and poppy are grown intensively in a process that involves the clearing of land, the planting of the crop, and its protection against pests including weeds, insects, and pathogens. Depending on the region, the clearing of the land for production purposes may have large and only slowly reversible effects on the environment. As for other forms of agricultural production, the clear-cutting of forests for the purposes of coca and poppy production reduces biodiversity, contributes to the release of greenhouse gases, increases the loss of soil nutrients, and promotes erosion of soils. Because production is illegal, it usually takes place in remote locations. As a result, clearing of land is done without government approval and with little apparent consideration for the biological and aesthetic value of the ecosystem.

A number of pesticides are used in the production of illicit drugs. Herbicides may be used in the initial clearing of the land and later in the suppression of weeds. Similarly, insecticides and fungicides may be used to protect the illicit crops from pests and diseases. To increase yields, fertilizers and other nutrients may also be used. Large quantities of agrochemicals have been seized and confiscated as part of the program to control the production of illicit drugs (Dirección Nacional de Estupefacientes 2002). Although some of these agrochemicals are highly toxic to mammals and may have significant environmental impacts, accurate information on the type of formulation used, amounts used, their frequency of use, and the conditions of their use is not available. Because of this, it was not possible to conduct a detailed human health and ecological risk assessment. However, the relevant toxicological and environmental properties of these substances are summarized in two separate reports, and several of these are significant potential hazards to human health and the environment (CICAD/OAS 2004c, 2005).

In addition to the use of agrochemicals in the production of coca and poppy, large amounts of chemicals are used in the processing of the raw product into refined cocaine and heroin. Processing of the illicit drugs is conducted in remote locations and in the absence of occupational health and environmental regulations and controls. During and after use, these substances may be released into the environment and have significant impacts on human and animal health and the ecosystem. The toxicological and environmental properties of these substances are summarized in a separate Tier 1 Hazard Assessment Report (CICAD/OAS 2004c). Some of these substances have potentially large environmental and human health hazards, and a subset of these is discussed in more detail in a Tier 2 Hazard Assessment Report (CICAD/OAS 2005).

The growing of coca and poppy and the distribution of cocaine and opium/heroin in Columbia have been the focus of a national control and eradication program starting in the 1970s. The program involves a number of departments and agencies of the Colombian Government and is coordinated by the Dirección Nacional de Estupefacientes (DNE), an agency of the Ministry of the Interior and Justice. The program has three main foci: the control of production of coca and poppy, the control of the processing, purification, and transport of the cocaine and heroin, and the seizure and forfeiture of the profits of illicit drug production (Dirección Nacional de Estupefacientes 2002).

The eradication program for illicit crops in Colombia is the responsibility of the Antinarcotics Directorate of the Colombian National Police (DIRAN-CNP), supported by data gathering from other nations such as those in North America and Europe. The DIRAN conducts regular flights with aircraft that spray coca and opium poppy crops with a herbicide. The DIRAN reviews satellite imagery and flies over growing regions on a regular basis to search for new coca and opium poppy growth and to generate estimates of the illicit crops through high-resolution low-altitude imagery and visual observation. The DIRAN selects the locations of the illicit crops that are to be sprayed with input from the DNE or the Government of Colombia's Plan Colombia Office.

Several concerns have been raised about the use of glyphosate and adjuvants in the control of coca and poppy plants. These concerns range from damage to other crops to adverse effects on the environment and human health. In response to this, the Government of Colombia appointed an independent environmental auditor who reviews the spray and no-spray areas with the DIRAN and regularly monitors the results of spraying through field checks and analysis of data from computerized spray records.

The objectives of the present assessment and review are to provide a science- and data-based study of the use of glyphosate in the eradication program with a key focus on the environment and human health, to collect data for use in the assessment, and to address specific concerns that have been raised. As with all risk assessments, we have followed a framework based on those used in other jurisdictions (NRC 1986; USEPA 1992, 1998). This framework consists of a Problem Formulation, Effects and Exposure Assessment, and Risk Characterization for both humans and the environment. In conducting this review, we used data from the peer-reviewed scientific literature, from government documents, and from studies specifically conducted to address data gaps.

II. Problem Formulation

Problem formulation is a key step in the risk assessment process and places the use of the substances being assessed into a local context. It is recognized that the growing of illicit crops such as coca and poppy, as well as the refining of the cocaine and heroin, involves considerable impacts on the environment through clearing of forests and the use of a number of substances for promoting crop growth and refining of the drugs. Although the identity of these substances is known, the type of formulation, the quantities used, and their manner of use is largely unknown and exposures in workers cannot be easily estimated. While the hazard of these substances is known (CICAD/OAS 2004c, 2005), the risks cannot be estimated, as the logistics of collecting the human and environmental exposure data are very difficult and not without other risks. Because of this, and as it was the initial mandate of the Panel, the focus of this risk assessment is on the use of glyphosate and adjuvants for control of the illicit crops. In this case, the locations and amounts of application are known with accuracy and environmental risk can be better estimated.

In humans, there are no specific biomarkers for exposure to glyphosate that can be used to estimate historical exposures. For logistical reasons, it was not possible to measure exposures resulting from eradication spraying directly in the field. For that reason, in epidemiology studies, indirect measures of exposures such as ecological studies, where the indicator variable or exposure is defined by eradication spraying and crops production patterns, must be used.

A. Stressor Characterization

The potential stressors in this risk assessment are glyphosate, its formulants, and adjuvants, such as surfactants, that are added to the spray formulation to modify its efficacy. The properties of glyphosate and these substances are described in the following sections.

Glyphosate

Glyphosate is the active ingredient of a number of herbicide formulations and is one of the most widely used pesticides on a global basis. Uses include agricultural, industrial, ornamental garden, and residential weed management. In agriculture, the use of glyphosate is increasing and use in soybeans is significantly greater since the introduction of glyphosate-tolerant crops (Wolfenbarger and Phifer 2000). According to the U.S. National Pesticide Use Database (USNPD 2006), use in soybean increased by 330% between 1992 and 1997 and by 460% between 1997 and 2002. Other agricultural uses for glyphosate-based products include its use by farmers as a routine step in preplanting field preparation. Nonagricultural users include public utilities, municipalities, and regional transportation departments where glyphosate is used for the control of weeds or noxious plants. The environmental and human health properties of glyphosate have been extensively reviewed (Giesy et al. 2000; Solomon and Thompson 2003; Williams et al. 2000) and by regulatory agencies (NRA 1996; USEPA 1993a, 1997, 1999; World Health

Organization International Program on Chemical Safety 1994). The following sections highlight key issues with regard to those properties of glyphosate that are fundamental to the assessment of risks associated with the coca and poppy eradication programs in Colombia.

Structure and Chemical Properties. The most common technical form of glyphosate is the isopropylamine salt (IPA), *N*-(phosphonomethyl) glycine isopropylamine salt (MW, 226.16; CAS number, 1071-83-6). The chemistry of glyphosate is important in determining its fate in the environment. Glyphosate (Fig. 1) is a weak organic acid comprising a glycine moiety and a phosphonomethyl moiety and closely resembles naturally occurring substances. Glyphosate is not chemically reactive, is not mobile in air or soils, does not have great biological persistence, and does not bioaccumulate or biomagnify through the food chain (CWQG 1999; Giesy et al. 2000; USEPA 1993a; Williams et al. 2000; World Health Organization International Program on Chemical Safety 1994).

Glyphosate is readily ionized and, as the anion, will be strongly adsorbed to organic matter in soils of normal pH. It thus has little mobility in soils and is rapidly removed from water by adsorption to sediments and suspended particulate matter (Giesy et al. 2000).

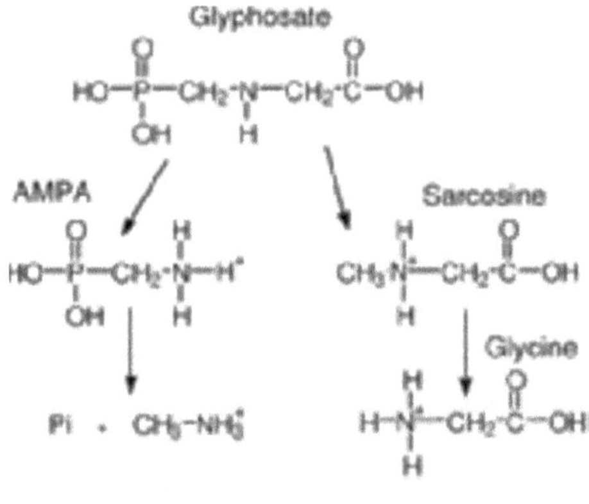

Fig. 1. Structure of glyphosate and its major metabolic and breakdown products.

Mechanism of Action. The mechanism of action of glyphosate is via the inhibition of the enzyme 5-enolpyruvyl shikimate-3-P synthetase, an essential enzyme on the pathway to the synthesis of the aromatic amino acids in plants (Devine et al. 1993; Franz et al. 1997). This inhibition results in decreases in the synthesis of the aromatic amino acids tryptophan, phenylalanine, and tyrosine, as well as decreased rates of synthesis of protein, indole acetic acid (a plant hormone), and chlorophyll. The death of the plant is slow and is first seen as a cessation of growth, followed by chlorosis and then necrosis of plant tissues. Inhibition of 5-enolpyruvyl shikimate-3-P synthetase is specific to plants. Many animals obtain their aromatic amino acids from plants and other sources and do not possess this pathway of synthesis. For this reason, glyphosate is relatively nontoxic to animals but is an effective herbicide in plants.

Global and Local Registration and Use. Glyphosate has been registered since 1971 and is currently widely used as a broad-spectrum, nonselective, postemergence herbicide in a number of countries around the world (World Health Organization International Program on Chemical Safety 1994). It is rapidly translocated from the leaves of treated plants to other parts of the plant, including the growing tips of stems and roots, and to underground storage organs, such as rhizomes and tubers. It is very effective for the control of perennial weeds and is more efficacious than many other nonselective herbicides that only affect the aboveground parts of the plant. Applied to soil, glyphosate shows little activity because the strong binding to soil organic matter makes it biologically unavailable for uptake by plants. Glyphosate has been used extensively in Colombia and many other countries for agricultural and other purposes for many years. Use of glyphosate in the coca and poppy spray program is shown in Table 1 and represents a relatively small fraction of the total use in Colombia.

Table 1. Glyphosate use in Eradication Spraying in Colombia, 2000–2004

Year	Amount sold in Colombia (L)[a]	Amount used in the eradication of illicit crops (L)[b]	Percent of total amount sold
2000	7,037,500	603,970	8.6%
2001	9,473,570	984,848	10.4%
2002	NA	1,061,538	11%[c]
2003		1,381,296	14%[c]
2004		1,420,130	14%[c]

[a]Data from ICA (2003).
[b]Data from Dirección Nacional de Estupefacientes (2002); Policia Nacional Dirección Antinarcóticos (2005).
[c]Estimated from total used in 2001 but likely less than this value.

Environmental Fate. The environmental fate of glyphosate has been extensively reviewed (CWQG 1999; Giesy et al. 2000; NRA 1996; World Health Organization International Program on Chemical Safety 1994); only key issues relevant to water and soil/sediment are summarized next.

As a result of its specific physicochemical properties, glyphosate is immobile or only slightly mobile in soil. The metabolite of glyphosate, aminomethyl phosphoric acid (AMPA; see Fig. 1), is somewhat more mobile in soil but is rapidly broken down, resulting in minimal amounts leaching in normal agricultural soils. The strong binding of glyphosate to soil results in almost immediate loss of biological activity; however, the bound residues do break down sufficiently rapidly that accumulation will not occur, even over many years of regular use. Contamination of groundwater from the normal use of glyphosate is unlikely except in the event of a substantial spill or other accidental and uncontrolled release of large amounts into the environment.

The high water solubility of glyphosate and its salts suggests that it would be mobile in water; however, strong and rapid binding to sediments and soil particles, especially in shallow, turbulent waters, or those carrying large loads of particulates, removes glyphosate from the water column (Tooby 1985). In normal agricultural uses, it is not expected to run off or leach into surface waters.

In water, the two major pathways of dissipation are microbiological breakdown and binding to sediments (Giesy et al. 2000; World Health Organization International Program on Chemical Safety 1994). Glyphosate does not degrade rapidly in sterile water, but in the presence of microflora (bacteria and fungi) in water, glyphosate is broken down to AMPA (see Fig. 1) and eventually to carbon dioxide (Rueppel et al. 1977). Other metabolic pathways have been reported (Liu et al. 1991), including further degradation of AMPA to inorganic phosphate and CH_3-NH_3, and via sarcosine to glycine (see Fig. 1). None of these products is considered herbicidal and would not be expected to be highly toxic to aquatic organisms at concentrations that would result from field use of glyphosate in aquatic systems. Photodegradation also may take place under field conditions where sufficient penetration of UV radiation occurs.

The dissipation of glyphosate from treated foliage and from leaf litter has also been characterized. As would be expected, most of the glyphosate sprayed on the plants penetrates into plant tissues after application, but some is available for washoff for several days after application (World Health Organization International Program on Chemical Safety 1994). If the plant dies as a result of this exposure, glyphosate would be present in the dead and decaying plant tissues. Glyphosate residues in leaf litter dissipate rapidly with a time-to-50%-disappearance (DT_{50}) of 8–9 d under temperate forestry conditions (Feng and Thompson 1990). Similar rapid dissipation from fruits and lichen has also been observed (Stiltanen et al. 1981) in north temperate regions.

Under tropical conditions such as in Colombia, dissipation will likely be more rapid than in temperate regions because of higher temperatures and moisture content, which promote microbiological activity as well as chemical degradation of many pesticides. Large areas of Brazil, Colombia, and Central America share similar tropical conditions and depend heavily on herbicides such as glyphosate (Racke et al. 1997). Glyphosate has been used in large areas of Brazil on no-tillage crops in general and, more recently, on transgenic soybeans. Comparing the fate of pesticides in tropical and temperate conditions, Racke et al. (1997) found no evidence of particular behavior of pesticides in the tropics and concluded a greater rate of degradation occurs under tropical conditions.

Formulants and Adjuvants

Formulants are substances that are added to a pesticide active ingredient at the time of manufacture to improve its efficacy and ease of use. These formulants serve many purposes and comprise a large range of substances, from solvents to surfactants to modifiers of pH. The glyphosate formulation used in Colombia includes several formulants. Adjuvants are added to formulated pesticides at the time of application and, like formulants, increase efficacy, or ease of use in special situations where pests are difficult to control or where nontarget effects need to be minimized. In the control program for illicit crops in Colombia, an adjuvant, Cosmo-Flux, is added at the time of spraying.

The relatively large water solubility and the ionic nature of glyphosate retard penetration through plant hydrophobic cuticular waxes. For this reason, glyphosate is commonly formulated with surfactants that decrease the surface tension of the solution and increase penetration into the tissues of the plants (Giesy et al. 2000; World Health Organization International Program on Chemical Safety 1994).

Surfactants in the Glyphosate Formulation. The glyphosate formulation as used in eradication spraying in Colombia contains formulants that are common to the commercial product as used in agriculture. Cosmo-Flux 411F, an agricultural adjuvant containing nonionic surfactants (a mixture of linear and aryl polyethoxylates, 17% w/v) and isoparaffins (83% v/v) (Cosmoagro 2004) is added to the spray solution. Adjuvants such as these are commonly added to pesticide formulations to improve efficacy through several mechanisms (Reeves 1992; Tadros 1994).

For example, surfactants such as the polyethoxylates in Cosmo-Flux, increase efficacy through increasing target surface adherence, promoting better droplet spread, better dispersion, prevention of aggregation, and enhanced penetration of herbicides into target plant tissues through the reduction of surface tension on plants. Surfactants can also disrupt the hydrophobic wax cuticle, thus increasing the penetration of active ingredient.

Base oils, such as the isoparaffins in Cosmo-Flux, are another class of adjuvants used in formulations. They are used primarily to aid foliar absorption of the pesticide by disrupting the waxy cuticle on the outer surface of foliage, which increases cell membrane permeability (Manthey and Nalewaja 1992).

Coca and Poppy Control Programs

The coca and poppy control programs make use of several procedures to identify, locate, and map coca and poppy fields. The initial step in this process is the use of satellite images to locate the fields. These images are provided by North American and European governments to the Government of Colombia. The images are used to locate potential areas of production. Further visual observations are made using overflights with observers and/or photographs from a low-altitude aerial photography aircraft, such as a Cessna Caravan, to verify the presence of coca and poppy fields. The camera used for this purpose is multispectral high resolution. Maps are generated in a Geographic Information System (GIS) and are used to produce updated coordinates for the spray pilots, as well as information for downloading into the aircraft navigation systems (Policia Nacional Dirección Antinarcóticos 2005). Field operation offices for the control program have computers and a satellite uplink for data transfer. Spray planes are equipped with high-resolution tracking equipment and Del Norte positional data recorders that display position, provide directional guidance, and store positional data and spray information on data cards for later analysis. Thus, field locations, flight paths of the spray planes, and areas where spray is released are known to within a resolution of 1–2 m.

Since 1994, coca and, more recently, poppy fields have been identified and sprayed. Total areas of identified fields and the area sprayed in Colombia are shown in Fig. 2. With increasing areas sprayed, the total area planted to coca has generally decreased since 2000.

Receiving Environment. Colombia is located between 4° S and 12° N of the equator. The country presents varied topography ranging from snow-capped peaks through high mountain plateaus to low-lying tropical regions. In general, coca tends to be grown at altitudes below 1,500 m and poppy at greater altitudes, usually 2,200 m. The biodiversity hotspot for the tropical Andean region includes significant areas of Colombia (Fig. 3). The tropical Andes biodiversity region is estimated to contain 15%–17% of the world's plant life in only 0.8% of its area. It has an area of 1,258,000 square kilometers, extends from Western Venezuela to Northern Chile and Argentina, and includes large portions of Colombia, Ecuador, Peru, and Bolivia (Centre for Biodiversity 2004).

Because the diversity hotspots are mainly associated with the Andean highlands and coca is mostly grown in lower altitudes, there is only some overlap between areas of coca production and regions of high biodiversity. Poppy is grown at a greater altitude, which overlaps with the biodiversity hotspot; however, the total areas grown at this time are small (see Fig. 2). Exact areas used for coca and poppy production within the diversity hotspot are not known; however, this information would be useful for assessing total impacts of production, especially for rare and endangered plant species.

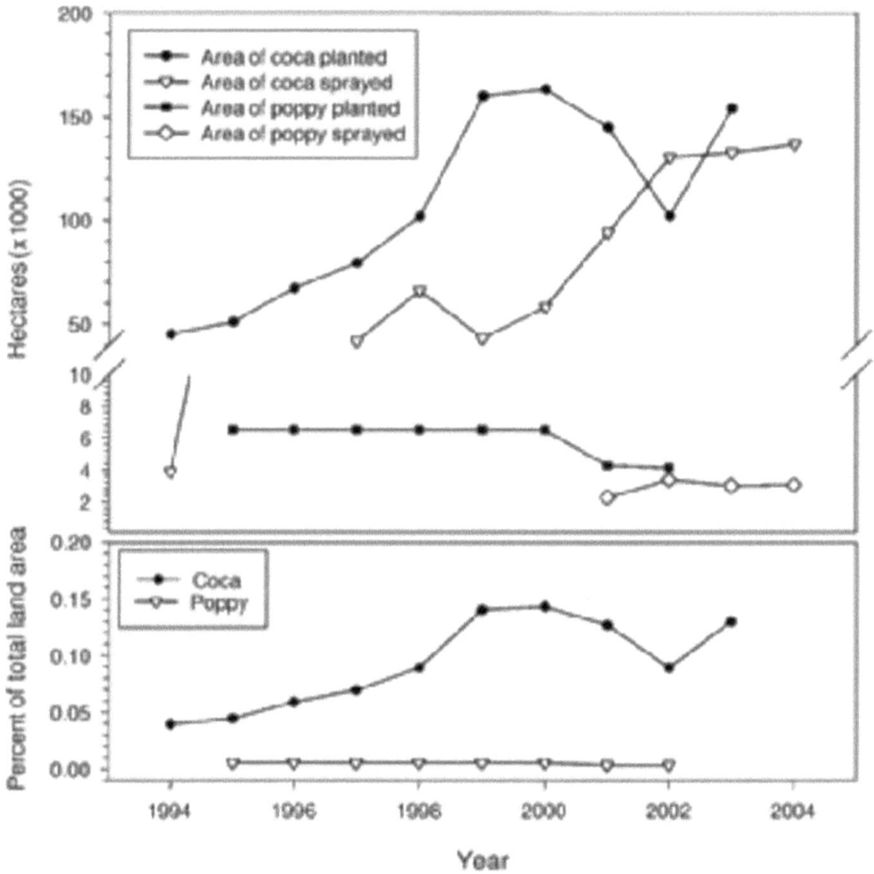

Fig. 2. Areas planted with coca and poppy in Colombia from 1994 to 2002 as hectares (ha) (*above*) and as a percent of the total land area of Colombia (*below*).

Method of Application. All coca and poppy fields are sprayed from fixed-wing aircraft. The procedure described below is based on observations recorded for the AT 65, AT 802, and OV 10 aircraft.

Spray planes are loaded in a special area of the tarmac (Fig. 4) at a number of bases throughout Colombia. Glyphosate and Cosmo-Flux are stored in plastic containers in a tarp-lined area protected by a berm to contain accidental spills. The areas may be in the open or covered. Glyphosate is transferred from 200-L plastic barrels to a larger plastic storage tank (Fig. 4A). Cosmo-Flux is transferred from 20-L plastic containers to a mixing tank. The required amounts of the components of the application mixture (glyphosate, Cosmo-Flux, and water from a local source) are pumped through a metering pump (Fig. 4B) into the aircraft using a table of mixing proportions to ensure the correct ratio of amounts are loaded.

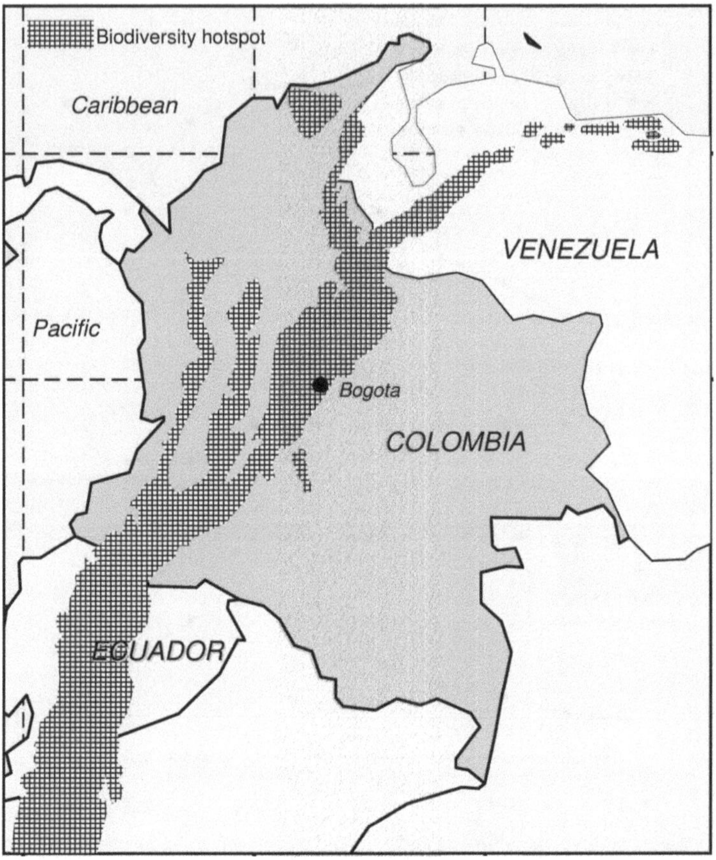

Fig. 3. Map showing the region of Colombia identified as part of the Andean Biodiversity Region.

Appropriate protective equipment is used by the mixer-loaders, who are trained in the loading procedures (Fig. 4C).

The spray boom (Fig. 4D) on the aircraft is equipped with raindrop nozzles (Fig. 4E). These nozzles produce droplets with a volume mean diameter (VMD) of 300–1,500 µm and are similar to those used in forestry spraying for site preparation (Payne 1993). The aircraft spray systems are electronically calibrated to disperse a specified quantity of spray mix per hectare, compensating for variances in ground speed. These electronic spray controls are checked each day by technicians and also during the pilot's preflight inspection. During actual spray operations, the pilot monitors the spray system by reading the spray pressure and flow rate gauges (United States Department of State 2002).

The same nozzles are used for both coca and poppy applications, but twice as many are used for the poppy applications at different boom

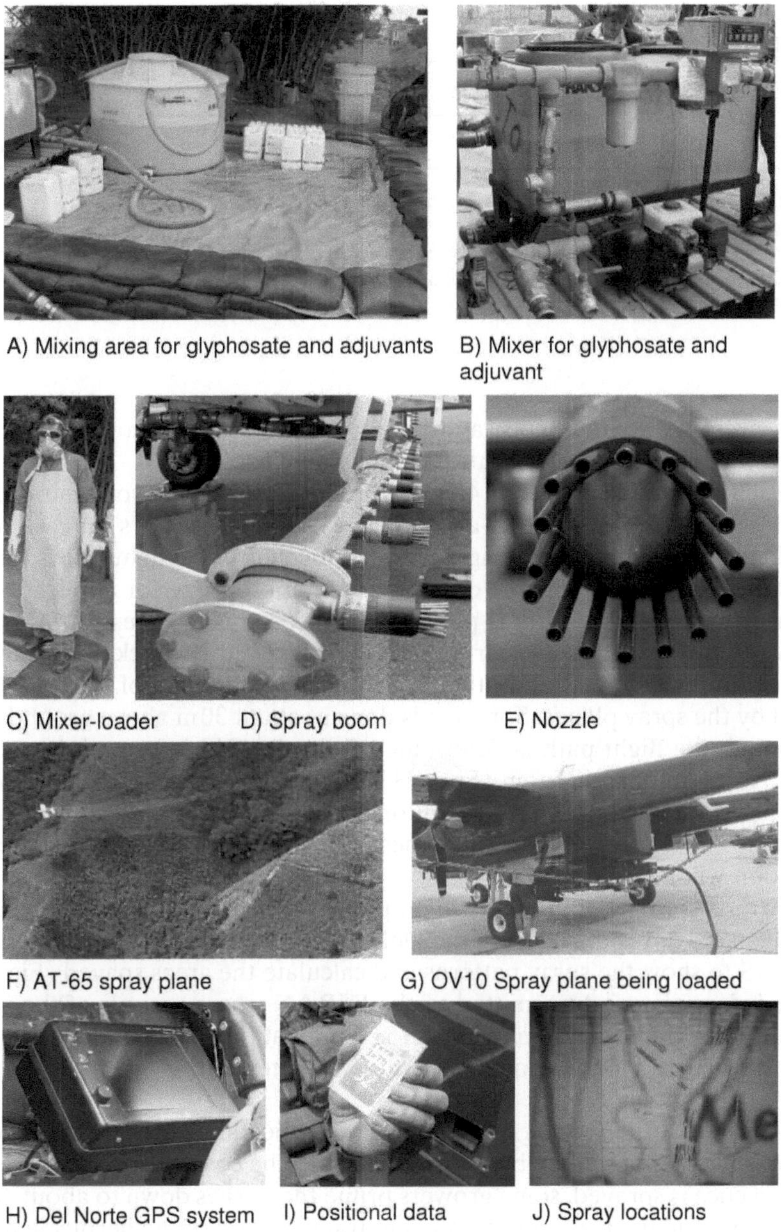

A) Mixing area for glyphosate and adjuvants
B) Mixer for glyphosate and adjuvant
C) Mixer-loader
D) Spray boom
E) Nozzle
F) AT-65 spray plane
G) OV10 Spray plane being loaded
H) Del Norte GPS system
I) Positional data
J) Spray locations

Fig. 4. Photographs of aspects of the spray operation.

Table 2. Application Rates of Glyphosate and Cosmo-Flux for Control of Coca and Poppy.

	L/ha		kg AE/ha[a]	
	Coca	Poppy	Coca	Poppy
Glyphosate	10.4	2.5	4.992	1.2
Cosmo-Flux	0.24	0.51		

[a]Glyphosate acid equivalent (AE)/ha. *Source*: Dirección Nacional de Estupefacientes (2002).

pressures. As a result, coca and poppy applications are done separately. Currently used application rates are shown in Table 2.

Each spray operation (Fig. 4F,G), which may consist of two or more spray planes, is escorted by search-and-rescue (SAR) helicopter(s) in case of an accident or incident. Spraying is only conducted in daylight hours before midafternoon to ensure that conditions are appropriate for application. If rain is imminent, visibility is poor, or wind speed is in excess of 7.5 km/hr (4 knots), spraying is not carried out. Wind speed is checked during the operation by the SAR and other helicopters with the aid of smoke generated by the spray planes. Spraying is done at about 30 m above ground and, although the flight path is determined from the GIS information and the Del Norte guidance system (Fig. 4H), the actual spraying is controlled by the pilots. In personal communications with five of the pilots, it was stated that, according to spraying guidelines, fields are not sprayed if people are or soon will be present.

After a spray operation, the flight path of the spray planes and the areas sprayed are downloaded from the Del Norte system (Fig. 4I) and processed by GIS to show the spray patterns and calculate the areas spayed (Fig. 4J). This information is transmitted to the DIRAN where records of the spray operations are retained and used for compilation of annual reports and statistics (Dirección Nacional de Estupefacientes 2002).

Frequency of Application. Frequency of application varies with local conditions and actions taken by the growers after the coca or poppy is sprayed. When coca is sprayed, some growers prune the bushes down to about 10 cm above ground in an attempt to prevent translocation of the herbicide to the roots. Sometimes these plants will recover and resprout; however, they will not yield large amounts of coca leaves for several months. If the field is replanted to coca from seedlings, reasonable productivity may not be achieved for 4–6 mon. If the field is replanted from cuttings, productivity may be achieved sooner. Thus, spraying of a particular coca field may have a return frequency of about 6–12 mon.

Being an annual, poppy is grown from seed. In the climatic conditions in Colombia, poppy fields would be harvested twice a year. If sprayed before reaching maturity and replanted immediately after spraying, they may be sprayed four times a year.

Exposure Pathways in Soil, Air, Water, and Other Media. In terms of application, there are several pathways through which glyphosate and adjuvants may contact the environment (Fig. 5).

Deposition on the target crop is the desired outcome; however, for purposes of assessing risks in humans and the environment, exposures that result in movement and deposition off the field are important. Spray drift would result in movement off target and could result in adverse effects in nontarget plants and animals. Given the strong adsorption of glyphosate to soil, deposition on soil in the field will likely not result in significant effects on nontarget organisms; however, runoff of residues bound to soil particles may result in contamination of surface waters with sediment-bound residues. Direct deposition and spray drift may result in contamination of local surface waters with glyphosate if these are in the spray swath or drift envelope of application. Depending on the depth of water, turbulence, flow, and presence of suspended particles, this would result in exposures of aquatic organisms to both glyphosate and any adjuvants present in the spray. Organisms present in the field during spraying would be exposed to the spray droplets and receive a theoretical dose, depending on surface area exposed and body mass.

Off-Target Deposition. There are two types of off-target deposition. The first is related to incorrect application where the spray pilot initiates application too soon or turns off the spray too late, or the spray swath includes a nontarget area on one or both sides of the target field. The second type

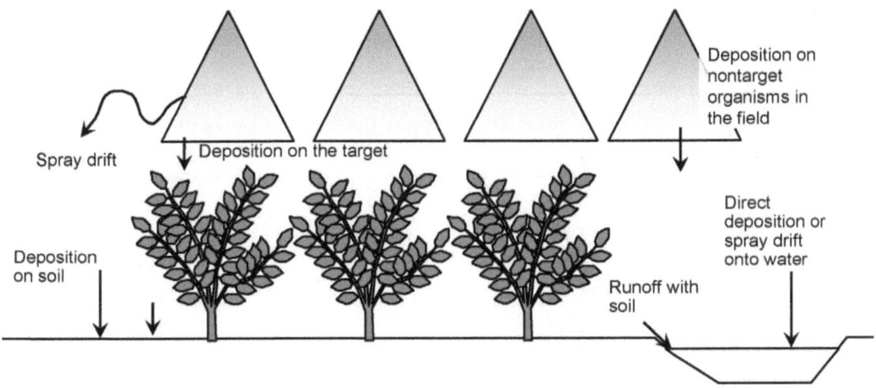

Fig. 5. Diagram showing exposure routes for various environmental compartments when glyphosate is used for the control of illicit crops.

of off-target deposition is spray drift. Experience with spray equipment of the type used in Colombia suggests that spray drift will be minimal (Payne et al. 1990). Estimates of accidental overspray have been made during assessments of spray program efficacy (Helling 2003). Based on site visits to 86 fields sprayed in 2002 and observations of damaged plants beyond the boundary of the area cleared and planted with coca, 22 fields showed evidence of off-field deposition. Using the size of these areas, it was estimated that between 0.25% and 0.48% of the areas cleared for coca production were damaged by offsite spray deposition (Helling 2003). Applying this estimate to the total area of coca sprayed (see Fig. 2) and calculating upper and lower intervals, the areas potentially affected are small when compared to the total area of Colombia (Table 3).

Although the areas affected by off-target drift are estimated to be small, this estimate is based on visual observations of a relatively small number of fields. These data were available only for coca, not poppy; however, the total areas planted to poppy are not large, and similar off-target deposition would be proportionately smaller than that associated with coca production. This lack of data is a source of uncertainty in the assessment. It is not logistically possible to visually inspect all sprayed fields; however, routine monitoring of the areas planted to coca and poppy that is undertaken by satellite and low-altitude imagery could be used to assess off-target deposition resulting in damage to plants. Changes in the size of sprayed fields over time could be used to extend these estimates over larger areas and increase their accuracy, although extension of the fields by growers may confound the data. The lower resolution of satellite imagery may preclude its use for this purpose; however, greater coverage by low-altitude images could facilitate this process.

Table 3. Estimates of Areas Affected by Off-Target Deposition of Glyphosate in the Spraying of Coca in Colombia.

Year	Hectares sprayed (ha)	Area affected by off-target deposits (ha)		Upper interval as percent (%) of the total area of Colombia
		Lower interval (0.25%)	Upper interval (0.48%)	
1994	3,871	9.7	18.6	0.0000002
1995	23,915	59.8	114.8	0.0000010
1997	41,861	104.7	200.9	0.0000018
1998	66,029	165.1	316.9	0.0000028
1999	43,111	107.8	206.9	0.0000018
2000	58,074	145.2	278.8	0.0000024
2001	94,152	235.4	451.9	0.0000040
2002	130,364	325.9	625.7	0.0000055
2003	132,817	332.0	637.5	0.0000056
2004	136,551	341.4	655.4	0.0000057

B. Framework for Risk Assessment

The following sections outline the conceptual model and hypotheses for the assessment of the human health and environmental impact of coca and poppy production in Colombia.

Conceptual Model

For purposes of the risk assessment of the use of glyphosate and adjuvants in the control of poppy and coca, the conceptual model applied was that normally applied to the agricultural application of pesticides where hazard and risk are directly related to toxicity and exposure. Thus, for human health, toxicity data were compared to exposures estimated from worst case data and also from more realistic data obtained in other uses of glyphosate, such as agriculture and forestry. Because of the low frequency of spray application, exposure from this source is acute and resulting risks were compared to acute toxicity data. Toxicity data for the active ingredient, glyphosate (IPA), were obtained from the literature and from the results of acute laboratory animal tests conducted with the mixture of formulated glyphosate and Cosmo-Flux as used in the spray program in Colombia. It is possible that glyphosate used in the eradication program may contribute to exposures via the food chain and drinking water; these exposures were estimated and compared with toxicity data and exposure guidelines based on chronic toxicity for glyphosate via dietary exposures. In addition, specific human health responses were assessed in epidemiological studies conducted in Colombia.

In assessing ecological risks, a similar agriculture-based approach was used. Similar to the foregoing approach, exposures were estimated from worst case models, from measurements made in other locations, and from measurements based on samples collected from the environment in Colombia. Because of the long periods between applications, ecological exposures from the spray operations are acute and were compared to acute toxicity data. Toxicity data were obtained from the literature and from laboratory-based tests on standard test organisms that were specifically conducted on the spray mixture as used in Colombia. The risk hypotheses are discussed next, and the remainder of this review focuses on tests of these hypotheses.

Risk Hypotheses

A large number of hypotheses were actually tested in this risk assessment; however, they were basically the same hypothesis with minor differences in the exposure and toxicity parameters. As is normal in the scientific method (Popper 1979), these hypotheses are stated as the null or negative hypothesis; thus, we attempted to falsify or disprove these hypotheses through the use of appropriate data.

For human health, two main hypotheses were used:

- Exposures to glyphosate and adjuvants as used in the poppy and coca eradication programs do not cause acute adverse effects to humans exposed via a number of routes.
- The use of glyphosate and adjuvants in those locations where eradication of poppy and coca are conducted does not result in acute and chronic health outcomes that are different from other locations where glyphosate is not used or is used in other agricultural practices.

For ecological effects, one main hypothesis was used:

- Exposures to glyphosate and adjuvants as used in the poppy and coca eradication programs do not cause acute adverse outcomes on nontarget organisms exposed via a number of routes.

III. Exposure Characterization

Exposure characterization is one of the key components of any risk assessment (NRC 1993; USEPA 1992, 1998). No measurements of farmer or pesticide applicator exposures have been made in Colombia. An assessment of pesticide use among farmers in the Amazon Basin of Ecuador has shown that paraquat and glyphosate are widely used. Risk behaviors were identified as frequent pesticide use, washing pesticide equipment in water sources used by humans, inadequate disposal of empty containers, eating and drinking during application, and using inadequate protective clothing (Hurtig et al. 2003). However, agricultural uses such as these are quite different from the aerial applications in Colombia. In the following sections, the potential for exposures in humans and the environment to glyphosate as used in the eradication program is discussed and characterized.

Human Exposure

In the agricultural setting, two groups are usually considered—applicators and bystanders. The group that experiences the greatest probability of exposure is the applicator group, which here includes mixer-loaders, spray plane pilots, and technicians who service the aircraft. The second group includes bystanders who may come into contact with the herbicide during application via direct deposition if they are within the spray swath, are directly exposed to spray drift, are exposed to deposits of spray when they reenter treated fields, or are exposed through the consumption of sprayed food items or contamined drinking water.

Applicator Exposure. Risk to applicators was not a specific target of this assessment; however, their exposure can be characterized. Based on observations of spray operations in several locations in Colombia, a number of measures are taken to reduce potential applicator exposure (Table 4).

Table 4. Protective measures used to reduce exposure of applicators to glyphosate and formulants as used in poppy and coca eradication programs.

Applicator subgroup	Mixer-loader	Spray pilot	Aircraft technician
Technology for handling of the formulation and spray mix	Use of closed-loading systems and pumps to mix and transfer glyphosate and Cosmo-Flux to the aircraft.	Not involved in mixing and loading	Not normally involved in mixing and loading; aircraft are washed down regularly so that exposure via contaminated surfaces is reduced
Protective equipment worn	Long pants, long sleeves, full rubber apron, rubber gloves, cloth hat or cap, particulate air filter and dark glasses, leather military-style boots	None other than normal clothing, long sleeves, long pants, jacket, and boots	Short or long sleeves, shorts or long pants, boots or sneakers, cloth cap or none
Equipment used to remove contamination, should it occur	Eye-wash station at all locations, clean water for washing hands and any contaminated surfaces, a shower in some locations	Same as is available to the mixer-loader	Same as is available to the mixer-loader

No measures of exposure were available for mixer-loaders in Colombia; however, they are likely to be similar to those of applicators in other situations. Based on observations on forestry and agricultural applicators (Acquavella et al. 2004; and summarized in Williams et al. 2000), exposures are generally small. Peak estimated exposure in applicators from all routes was 0.056 mg/kg body weight (bw). The estimate of chronic exposure from all routes was 0.0085 mg/kg/d based on an 8-hr day and a 5-d work week. In the results of the recently published Farm Family Exposure Study, the greatest estimated systemic dose in a sample of 48 applicators was 0.004 mg/kg bw (Acquavella et al. 2004). In Colombia, mixing and loading are done by one or two individuals wearing appropriate protective equipment. Pilots have limited opportunity for exposure and, as has been observed in other studies (Frank et al. 1985), likely experience less exposure.

Exposures of mixer-loaders in Colombia are likely to be similar to those observed in agricultural applications. Exposures for spray pilots and technicians will likely also be less than for applicators. While most of the protective clothing worn by mixer-loaders is appropriate, the need for a respirator is questionable and the use of dark glasses in place of a full face shield is judged inappropriate. Dark glasses will not protect the eyes from a splash to the forehead that runs into the eyes, a vulnerable area in terms of glyphosate exposure during mixing and loading (Acquavella et al. 1999). A full face shield offers better protection. As glyphosate is not volatile, nor atomized during mixing and loading, use of a respirator offers little reduction in potential exposure and complicates the use of a full face shield. The usefulness of a respirator is judged to be small.

Bystander Exposure. Bystanders can be classified into several classes, depending on their route of exposure. These are discussed in the following sections.

Bystanders Directly Oversprayed. Although it is unusual for people to be present in a coca field during application, it is possible that a person could be standing directly in the spray swath and would receive a direct application of the spray. Several scenarios could occur (Fig. 6, Table 5). The most likely is the partially clothed human with a cross-sectional area of 0.25 m^2 exposed to the spray (Table 5). Given that glyphosate penetrates poorly through the skin with maximum penetration of about 2% (Williams et al. 2000), the body dose under a reasonable worst case exposure will be approximately 0.08 mg/kg bw.

Bystander exposure to glyphosate was estimated as 0.0044 mg/kg bw/d for a child 1–6 yr of age (Williams et al. 2000). Exposures to glyphosate were measured in bystanders to farm applications (Acquavella et al. 2004). These studies were conducted in spouses and children not involved in applications, and frequency of measurable exposure was small, with 4% and 12% of the spouses and children, respectively, with detectable exposures based on

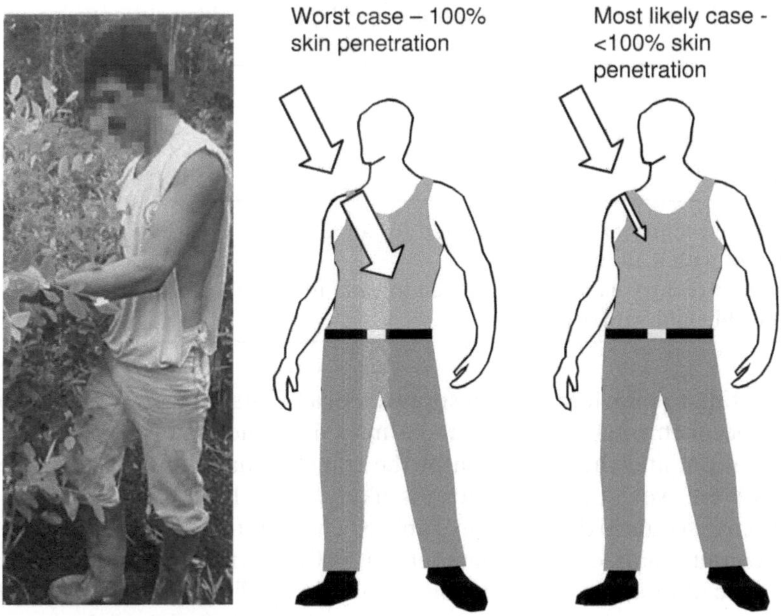

Fig. 6. Illustration of human exposure scenarios.

Table 5. Estimates of human exposure to glyphosate during a spray application.

	Exposure in mg/kg bw	
Scenario	Coca: 4.992 kg/ha	Poppy: 1.2 kg/ha
Partially clothed human with cross-sectional area of $0.25\,m^2$, complete penetration	1.8	0.4
Partially clothed human with cross-sectional area of $0.25\,m^2$, 2% penetration (most likely)	0.04	0.01

Assumptions: body weight, 70 kg; body surface area, $2\,m^2$.

urinary monitoring. The maximum systemic dose estimates for spouses and children were 0.00004 mg/kg and 0.0008 mg/kg bw, respectively (Acquavella et al. 2004). If bystanders are neither directly sprayed nor reenter the field immediately after spraying, their exposures will likely be within a factor of 10 of farm bystanders. All these measured exposures are considerably less than those estimated in Table 6, considered to be reasonable worst case values.

Reentry. If a person were to reenter the sprayed field immediately after spraying and come into contact with the treated foliage, such as when

Table 6. Estimates of human exposure to glyphosate during reentry to treated fields.

Scenario	Exposure in mg/kg bw	
	Coca: 4.992 kg/ha	Poppy: 1.2 kg/ha
Maximum reentry exposure estimated for an adult human with a 10-hr d	0.013	0.003
Maximum reentry exposure estimated for a 1- to 6-yr-old child with a 10-hr d	0.259	0.062

attempting to pick leaves from sprayed coca plants, exposure to glyphosate could occur through the hands and arms. Given the area exposed, the small penetration, and the saturation of the transfer that would result once the hands were wet, total body dose is likely to be less than the reasonable worst case scenario described in Table 6. The potential for reentry exposure has been summarized by Williams et al. (2000). Reentry exposures decreased with time after application and, on day 7 after application, were 3% of those estimated for day 1. Reentry into areas of tall weeds (height, 1.5 m) resulted in 10-fold-greater exposures than in areas of short grass. Based on measurements in farmworkers, estimates of reentry exposure to glyphosate in adults ranged from 0.0000039 to 0.0026 mg/kg bw/hr of reentry time. Maximum reentry exposure for a 1- to 6-yr-old child was estimated at 0.026 mg/kg bw for a 5-hr contact period. As these estimates are based on a spray application rate of 1 kg/ha, reentry exposures under Colombian conditions are estimated to be somewhat greater (see Table 6). These numbers are also greater than the direct overspray, as the persons involved may have repeated exposures if they reenter a field immediately after spraying.

Inhalation. Because the vapor pressure of glyphosate (isopropyl ammonium salt) is low (2.1×10^{-3} mPa at 25°C) and it also has a small Henry's law constant (4.6×10^{-10} Pa m^3 mol^{-1}) (BCPC 2003), it will not be present in air as a vapor at biologically relevant concentrations. The droplet sizes resulting from the spray application of glyphosate in Colombia are large, with a mean droplet diameter ~1,000 μm and with very few droplets <500 μm. As such, they are unlikely to be inhaled and penetrate into the lungs. Based on measurements of glyphosate concentrations in air during applications, the maximum estimated daily dose (8 hr) resulting from inhalation of spray droplets by applicators was 0.0062 mg/kg bw (Williams et al. 2000), a value that is judged to be applicable as a maximum exposure for bystanders.

Dietary and Drinking Water. Dietary and drinking water exposures to glyphosate have been estimated to be relatively small under conditions of use in North America (Williams et al. 2000) (Table 7).

Table 7. Worst case daily human exposure estimates for glyphosate (mg/kg bw/d)

	Female adult		Female child (1–6 yr)	
Sources	Acute	Chronic	Acute	Chronic
Drinking water	0.000036	0.000002	0.000110	0.000004
Diet	0.024	0.024	0.052	0.052
Wild foods	0.045		0.045	
Total from diet and water	0.069	0.024	0.097	0.052
Values extrapolated from the above (Williams et al. 2000) to the greater application rate of 4.992 kg/ha used in control of coca				
Drinking water	0.000179	0.00001	0.00055	0.000018
Diet	0.119	0.119	0.259	0.259
Wild foods	0.224	0.224	0.224	0.489
Total from diet and water	0.343	0.293	0.483	0.747

The results of monitoring programs conducted by the Danish Veterinary and Food Administration from 1997 to 1999 reported on the content of glyphosate and several other pesticides in cereals produced in Denmark (Granby and Vahl 2001). Based on the residues of glyphosate in cereals, intake of glyphosate for a 60-kg adult was estimated at 0.007 mg/d.

Based on a study of 51 streams in nine midwestern U.S. states, the U.S. Geological Survey (USGS) reported the presence of glyphosate and a number of other herbicides in surface waters (Battaglin et al. 2005; Scribner et al. 2003). Of a total of 154 water samples collected during 2002, glyphosate was detected in 36%, and its degradation product, aminomethylphosphonic acid (AMPA), was detected in 69%. The greatest measured concentration of glyphosate in any sample was 8.7 µg/L, and the greatest concentration of AMPA detected in the USGS study was 3.6 µg/L. More recently, glyphosate and AMPA have been detected in association with urban inputs from wastewater treatment in the U.S. Maximum concentrations of glyphosate and AMPA were 2 and 4 µg/L, respectively (Kolpin et al. 2006). Concentrations of glyphosate detected in Colombian surface waters (see following) were usually less than 25 µg/L, the detection limit. Exposures from drinking untreated surface waters in areas where spraying takes place are judged to be small and infrequent.

Environmental Exposures

Air. The presence of glyphosate in air is unlikely because it, and the salt forms commonly used in glyphosate formulations, have essentially negligible vapor pressure. Spray droplets may, however, be present in air and are the likely reason for detection of glyphosate, along with other pesticides, in rainwater in the European Union (EU) (Quaghebeur et al. 2004). From 1997 to 2001, glyphosate was only detected in rainwater in Belgium in 2001 and then with a frequency of 10% and a maximum concentration of 6.2 μg/L.

Water. If water is directly oversprayed, contamination of surface waters will result (see Fig. 5). Some coca fields are located near ponds and lakes and some are near streams and rivers (Helling 2003). Although surface waters are not deliberately sprayed, some overspray of small watercourses and the edges of ponds, reservoirs, and lakes may occur. In the absence of measured concentrations immediately after spraying in surface waters located close to fields, estimates of exposure were made using worst case assumptions (Table 8) based on water depth assumptions used by the USEPA (Urban and Cook 1986) and the EU (Riley et al. 1991).

Glyphosate has been detected in surface waters (see foregoing discussion on human exposures through drinking water) in a number of locations.

Table 8. Estimates of concentrations of glyphosate in surface water after a spray application.

	Exposure in μg/L (glyphosate[a])	
Scenario	Coca: 4.992 kg/ha (3.69 kg AE/ha)	Poppy: 1.2 kg/ha (0.89 kg AE/ha)
Surface water, 2 m deep, rapid mixing and no absorption to sediments, no flow	185	44
Surface water, 0.3 m deep, rapid mixing and no absorption to sediments, no flow	1,229	296
Surface water, 0.15 m deep, rapid mixing and no absorption to sediments, no flow	2,473	595
Surface water, 0.15 m deep, rapid mixing and 50% absorption to sediments, no flow	1,237	297

[a]Note that the concentration is expressed as glyphosate acid (AE) to allow comparison to exposures used in environmental toxicity testing. In both these exposures and in the toxicity testing of Cosmo-Flux, proportional amounts are present and the exposure and toxicity values are thus directly comparable and can be used to assess the hazard of the mixture as applied in Colombia.

Glyphosate residues have been reported in surface waters in Denmark as a result of agricultural activities. These residues were observed as part of the Pesticide Leaching Assessment Program (PLAP), a project that was intended to study leaching potential of pesticides to groundwater (Kjaer et al. 2003, 2005). PLAP was focused on pesticides used in farming and monitored leaching at six agricultural test sites representative of Danish conditions. Water from special drilled wells and from normal tile drains was analyzed for glyphosate and aminomethylphosphonic acid (AMPA, a major degradate of glyphosate). It is not clear from the report if the samples were filtered before analysis; this is important as glyphosate binds strongly to organic matter in soils and can be transported in this form. The presence of macropores in soil would facilitate transport to the tile drains.

In the samples from PLAP collected following glyphosate applications, there were no detections of glyphosate or AMPA that exceeded 0.1 μg/L in any of the groundwater samples taken from suction cells (1 and 2 m below the surface), the vertical wells (about 1.5–5.5 m below surface), and the horizontal wells (about 3.5 m below surface).

Glyphosate residues were detected in water from tiles draining the field and were observed primarily in the autumn. The highest measured concentrations were 5.1 μg/L for glyphosate and 5.4 μg/L for AMPA. The calculated average annual concentrations of glyphosate and AMPA in drainage water were 0.54 and 0.17 μg/L, respectively, at one location, and 0.12 and 0.06 μg/L, respectively, at a second. At a third location, glyphosate and AMPA were detected but average concentrations of both were below 0.1 μg/L. In Danish soils, degradation of glyphosate was shown to be slower in sandy soils than gravel but leaching was observed only in rounded gravel soils (Strange-Hansen et al. 2004), and leachate concentrations were less than 0.1 μg/L (Fomsgaard et al. 2003). Similarly, a study on fate of glyphosate in soils showed rapid dissipation with almost total dissipation 1 mon after application (Veiga et al. 2001). Given the small organic content of gravel and the presence of macropores between the grains of gravel, movement through this matrix is not surprising. Complete degradation in other types of soil is expected.

Other authors have reported glyphosate residues in surface waters in Europe (Skark et al. 1998, 2004), although the frequency of detection was not large. These authors suggested that the contamination was from application to railroad beds, environments where gravel is used and where adsorption would be expected to be minimal. This conclusion is supported by other studies on the dissipation of herbicides applied to railroad beds (Ramwell et al. 2004) and highways (Huang et al. 2004; Ramwell et al. 2002). Application of glyphosate to hard surfaces in an urban context (road edges) can give peak runoff concentrations of 650 μg/L (Ramwell et al. 2002), but only 15 μg/L from a railway trackbed (Ramwell et al. 2004). In Germany, a study of two catchments found that nonagricultural pesticide use contributed more than two-thirds of the whole observed pesticide load in the tributaries and at least one-third in the Ruhr River (Skark et al. 2004). Most nonagricultural pesticides were derived from runoff from domestic, industrial and railway areas. Nevertheless, in Argentina, where glyphosate-tolerant soybean is now extensively grown and regularly treated, no residues have been observed in soil or water of either glyphosate or AMPA (Arregui et al. 2004).

The USGS study on midwestern U.S. streams (Battaglin et al. 2005; Scribner et al. 2003) analyzed water samples filtered through a 0.7-μm filter; thus, the concentrations represent dissolved glyphosate and AMPA. Measured values in this study ranged up to 8.7 μg/L.

Although the glyphosate concentrations in surface waters in other areas where it is used in agricultural and other activities are relatively small, concentrations have not been measured in Colombia. To address this uncertainty, we conducted a monitoring study to measure levels of glyphosate, AMPA, and other pesticides in surface waters.

This study was conducted in five locations in Colombia representing areas where spraying of coca was planned or where other agricultural activities were undertaken and were also near the human health studies. Sites were selected for safe access as well as ease of repeated sampling and are summarized in Table 9 with further details of temperatures, rainfall, and soil characteristics from separate reports (PTG 2005a–e).

To characterize concentrations of glyphosate and AMPA in surface waters, samples were taken weekly for 24 wk (CICAD/OAS 2004a). Samples, in plastic bottles, were frozen and held at −17°C until shipped to Canada for analysis using published methods (Thompson et al. 2004). The method detection limit (MDL) for the analysis was 25 μg/L. Duplicate samples were taken and one sample held in Colombia until the duplicate had been analyzed. In addition, field-spiked samples and blanks were taken at biweekly intervals. In addition to water, sediment samples were taken at monthly intervals for analysis of glyphosate and AMPA if significant concentrations were detected in surface waters. Appropriate field spikes and blanks of sediment were also taken bimonthly. Quality control samples showed excellent recovery and precision of the analytical method with 98% recovery for glyphosate and 8.8% coefficient of variation (CV), and 110% recovery efficiency for AMPA with 20% CV. Blank field sample analyses show no coextractive interferences above the MDL for either glyphosate or AMPA at any of the sample sites. Field-spiked samples showed no significant degradation of glyphosate during handling and transport with overall average value of 90% of expected concentrations.

Results are summarized in Table 9. In all locations and on most occasions, residues of glyphosate and AMPA were present at concentrations below the MDL of 25 μg/L. On one occasion each in Valle del Cauca and Boyacá, glyphosate concentrations of 30.1 and 25.5 μg/L, respectively, were found.

Table 9. Characteristics of sampling sites for glyphosate and other pesticides in surface waters and sediments in regions of Colombia and measured values in samples collected between October 2004 and March 2005.

Site name	Location	Altitude (m)	Major crop types	Known pesticide use	Frequency of detection of glyphosate (*n* of 24, %)	Other pesticides detected (*n*/*N*) and types
Valle del Cauca, Río Bolo	3°27.642′ N 76°19.860′ W	1002	Sugar cane	Glyphosate and other pesticides	1 (4%)	(3/10) 2,4-D
Boyacá, Quebrada Paunera	5°40.369′ N 74°00.986′ W	557	Coca	Manual eradication, no aerial spraying of glyphosate	1 (4%)	(0/8)
Sierra Nevada, Quebrada La Otra	11°13.991′ N 74°01.588′ W	407	Organic coffee	None	0 (0%)	(0/9)
Putumayo, Río Mansoya	0°43.259′ N 76°05.634′ W	329	Coca	Aerial eradication spraying	0 (0%)	(0/9)
Nariño, Rio Sabaletas	1°27.915′ N 78°38.975′ W	15	Coca	Aerial eradication spraying	0 (0%)	(1/8) endosulfan I, endosulfan II, endosulfan sulfate

At these sites, spraying was not carried out and the only use of glyphosate, if any, was in agriculture. These data suggest that, at the watershed level, little or no contamination of surface waters with glyphosate at significant concentrations has resulted from the use of glyphosate in either agricultural or eradication spraying in Colombia. As concentrations in surface waters were mostly below the MDL, sediment analyses were not performed.

To characterize concentrations of other pesticides in surface waters and sediments, samples of water were taken in glass bottles every 2 wk for 22 wk (CICAD/OAS 2004b). Samples were held at 4°C until shipment to Canada for analysis. Analyses were conducted at the Laboratory Services Division of the University of Guelph using standard methods (LSD 2005). Duplicate samples were held in Colombia until analyses were completed. Field spikes and blanks were taken at 5-wk intervals, as were sediment samples. Sediment blanks and spikes were taken only once. These results are also summarized in Table 9. Blanks showed no contamination of samples during storage and shipping. Spiked samples showed variable recovery, particularly for the carbamate, carbaryl. Several pesticides were detected in surface waters, which is not unexpected as pesticides are widely used in agriculture in Colombia and, based on experience in other locations, some contamination of surface waters will occur. Of interest is the detection of endosulfan (I and II) and its breakdown product, endosulfan sulfate, in samples taken at the Nariño site. Endosulfan is not registered for use in Colombia, and its detection here likely is the result of illegal use. Whether this contamination resulted from regular agricultural activity or from use in the production of coca is unknown.

Soil. Concentrations of glyphosate in the top 25 mm of soil were estimated from the application rates and ranged from 1.6 to 3.2 mg/kg for poppy use rates and from 6.7 to 13.3 mg/kg for coca, depending on assumptions about interception by the crop foliage (50%) and soil density (1.5 kg/L). Measurements could be made through the use of residue analysis; however, the more important question is the biological availability of the glyphosate, as this would determine its potential for biological effects.

Although there are no direct measurements of glyphosate and AMPA concentrations available from treated coca and poppy fields in Colombia, the biological activity of any residues that may be present is judged to be small as the sprayed fields rapidly become colonized with invasive plants or are replanted to coca soon after spraying. From visual observations (Fig. 7), from observation in other uses and other locations (above), and from other reports (Helling 2003), recolonization is rapid and there have been no adverse effects observed in terms of recolonization or replanting of sprayed fields.

Fig. 7. Photograph of coca plants near Caucasia, Colombia, replanted from cuttings in a field sprayed with glyphosate 56 days previously.

IV. Effects Characterization

A. Glyphosate

Human health and environmental effects of glyphosate have been extensively reviewed (Giesy et al. 2000; Solomon and Thompson 2003; Williams et al. 2000) and by regulatory agencies (NRA 1996; USEPA 1993a, 1997, 1999; World Health Organization International Program on Chemical Safety 1994). The following sections are primarily directed to a critical analysis of original articles published since 1999 or that were not included in the earlier reviews (Giesy et al. 2000; Solomon and Thompson 2003; Williams et al. 2000). In characterizing the effects of glyphosate, it is important to distinguish between glyphosate as the active ingredient (usually glyphosate IPA salt) and the formulated product, such as Roundup. Glyphosate salts readily dissociate into the free acid, and the acid and salts are considered toxicologically equivalent. Formulations of glyphosate contain additional formulants that modify uptake of the glyphosate into plants and may alter toxicity of the mixture. In the following sections, tests conducted with the active ingredient only are referred to as "glyphosate." Those tests where a formulation was used are referred to by the specific product name, or where this is not known, as "glyphosate formulation." To allow easy comparison between technical product and formulations, where possible, concentrations of glyphosate have been normalized to acid equivalents (AE).

Effects of Glyphosate on Mammals

Laboratory Toxicity Studies. The toxicity of glyphosate and the formulation Roundup were reviewed by Williams et al. (2000). Glyphosate acid and its isopropylamine salt have little acute toxicity by the oral, dermal, and subcutaneous routes of exposure (Table 10).

Toxicity was greatest by intraperitoneal administration. When rats and mice were given glyphosate orally or intraperitoneally, several stress symptoms, such as increased respiration, elevated rectal temperatures, and occasional asphyxial convulsions, were noted. Median lethal doses of 4,704 mg/kg bw to the rat and 1,581 mg/kg bw to the mouse orally were significantly higher than 235 and 130 mg/kg bw, respectively, median lethal doses obtained when glyphosate was given intraperitoneally. Lung hyperemia was the major lesion noted in the glyphosate-poisoned animal (Bababurmi et al. 1978).

There is limited information on acute toxicity in dogs. However, there is a retrospective study conducted of 482 glyphosate-related calls recorded at the Centre National d'Informations Toxicologiques Vétérinaires (CNITV) of France between 1991 and 1994. Only 31 cases were assessed as certain or highly probable and were linked with direct ingestion of

Table 10. Acute toxicity of glyphosate and formulations in selected mammals.

Species	Route	Compound administered[a]	LD_{50} (mg/kg bw)
Mouse	Oral	Glyphosate	>10,000
		Glyphosate	1,538
	Subcutaneous	Glyphosate saline	6,250 (M)
		Glyphosate saline	7,810 (F)
	Intraperitoneal	Glyphosate saline	545 (M)
		Glyphosate saline	740 (F)
		Glyphosate	134
Rat	Oral	Glyphosate, Roundup, glyphosate isopropylamine salt	>5,000
	Dermal	Roundup	>17,000
	Inhalation	Roundup, glyphosate saline	LC_{50} = 3.18 mg/L (4 hr)
	Subcutaneous	Glyphosate saline	17,500
		Glyphosate saline	281 (M)
			467 (F)
	Intraperitoneal	Glyphosate	238
Rabbit	Oral	Glyphosate	3,800
	Dermal	Glyphosate, Roundup, glyphosate isopropylamine salt	>5,000
Goat	Oral	Glyphosate, Roundup, glyphosate isopropylamine salt	>3,500

Source: Smith and Oehme (1992).

glyphosate concentrates or spray in 25 dogs. The symptoms were most frequently described as vomiting, hypersalivation, and diarrhea; prostration and paresis were not common. Symptomatic treatment resulted in rapid recovery without sequelae (Burgat et al. 1998). Campbell and Chapman (2000) described the onset of clinical effects in dogs observed in several cases of poisoning as usually between 30 min and 2 hr. Recovery usually occurs over 1–2 d. Salivation, vomiting, diarrhea, irritation, and swelling of lips are common early features. Tachycardia and excitability are often present in the early stages, with the animals subsequently becoming ataxic, depressed, and bradycardic. Inappetence, pharyngitis, pyrexia, twitching, shaking, and dilated pupils are noted occasionally. Rarely, jaundice, hepatic damage, and hematuria have been reported. Eye and skin irritation are also possible. Tachypnoea occurs in glyphosate poisoning in other animals but does not appear to be a feature of glyphosate toxicity in dogs.

Studies to examine the effects of chronic feeding of glyphosate to Wistar rats have measured the activity of some enzymes with a function in the pathways of NADPH generation, isocitrate dehydrogenase, glucose-6-phosphate dehydrogenase, and malate dehydrogenase in liver, heart, and brain of pregnant Wistar rats and their fetuses that were exposed to glyphosate solutions of 0.5% and 1% at a dose of 0.2 and 0.4 mL/mL water during 21 d of pregnancy. Glyphosate affects these enzymes in the studied organs of pregnant rats and their fetuses (Daruich et al. 2001).

Feeding Glyphosate-Biocarbo formulation at rates of 4.87 mg/kg every 2 d for 75 d resulted in leakage of the hepatic intracellular enzymes alanine aminotransferase (ALT) and aspartate aminotransferase (AST), suggesting irreversible damage in hepatocytes (Benedetti et al. 2004). The formulation used in this study was from Brazil, and the identity and composition of the formulants are unknown. In addition, the exposures extended over a long period and were judged inappropriate for assessing risks from acute and infrequent exposures such as may occur in eradication spraying.

The effect of glyphosate on the activity of several enzymes was studied *in vitro*. The enzymes measured were serum acetylcholinesterase (AChE), lactate dehydrogenase (LDH), aspartate amino-transferase (AST), alanine aminotransferase (ALT), alkaline phosphatase (AP), and acid phosphatase (AcP). Glyphosate inhibited all enzymes except AcP. IC_{50} values were 714.3, 750, 54.2, 270.8, and 71.4 mM for ACHE, LDH, AST, ALT, and AP, respectively (El-Demerdash et al. 2001). The most sensitive response, that of AST, was observed at 54.2 mM, equivalent to a concentration of 9,056 mg/L, a concentration that would not occur *in vivo*. These results do not suggest that glyphosate would have effects at concentrations lower than those previously observed.

Glyphosate has not been found to be mutagenic, genotoxic, or carcinogenic. Glyphosate was not teratogenic or developmentally toxic except at large exposures (Williams et al. 2000). Some studies that were not reviewed by Williams et al. or were published after 2000 are reviewed below.

In a study on Charles River CD-1 rats, test animals were given oral gavage doses of 0, 300, 1,000 and 3,500 mg/kg bw/d of glyphosate from day 6–19 of gestation. Control animals received 0.5% methocel. No internal or skeletal anomalies were seen at 300 and 1,000 mg/kg bw/d, although maternal toxicity was apparent at 3,500 mg/kg bw/d with soft stools, diarrhea, red nasal discharge, reduced body weight, and death by gestation day 17 (6/25). In addition, mean fetal body weights were significantly reduced and early fetal resorption was significantly increased at this dose (Rodwell 1980b). Female Dutch belted rabbits were given oral gavage doses of 0, 75, 175, and 350 mg/kg bw/d glyphosate from day 6–27 of gestation. Control animals received 0.5% methocel. No internal or skeletal abnormalities were seen (Rodwell 1980a). In a study from Brazil, examination of pregnant Wistar rats dosed orally with Roundup from day 6–15 of pregnancy with rates of 0, 500, 750, or 1,000 mg/kg bw glyphosate showed skeletal alteration in fetuses (15.4%, 33.1%, 42.0%, and 57.3%, respectively). There was 50% mortality of dams at 1,000 mg/kg only (Dallegrave et al. 2003). The doses were large and considerably greater than those used in an earlier study (reviewed by Williams et al. 2000). In the earlier study, a no-observed-effect-level (NOEL) of 15 mg/kg bw/d was described for fetal effects and 300 mg/kg bw/d for maternal effects. Given the very large doses used in the Dallegrave et al. (2003) and Rodwell studies (1980), their results are not surprising and do not change the assessment of teratogenic potential in Williams et al. (2000).

A number of recent studies have been carried out in tissue culture. One assessed the affect of several formulated pesticides on the steroidogenesis pathway (StAR protein synthesis) in tissue cultures of mouse testicular Leydig tumor cells (Walsh et al. 2000). Exposure to the formulation at 25 mg/L in the cell culture medium caused a reduction in steroidogenesis but only for a period less than 24 hr during which there was recovery. In another study, Lin and Garry reported results of bioassays carried out in cultures of the MCF-7 breast cancer cell (Lin and Garry 2000). Results indicated that although some pesticides caused estrogen-like receptor-mediated effects at large exposure concentrations, both glyphosate and the Roundup formulation induced nonestrogen-like proliferation, thereby supporting the view expressed by others (Williams et al. 2000) that neither glyphosate nor Roundup is an endocrine disruptor.

Studies on cells *in vitro* are difficult to interpret as they exclude the normal pharmacokinetic and metabolic functions that would be present in whole animals; thus, these should be compared to the multigenerational study used by regulatory agencies worldwide to assess reproductive/developmental toxicity, which is the most definitive study design for the evaluation of potential endocrine modulating substances in humans and other mammals. Comprehensive reproductive and developmental toxicology studies carried out in accordance with internationally accepted protocols have demonstrated that glyphosate is not a developmental or reproductive toxicant and is not an endocrine disruptor (Williams et al. 2000; USEPA 1993a; World Health Organization International Program on Chemical Safety 1994).

There was no evidence of neurotoxicity in a number of studies reviewed in Williams et al. (2000). Neurotoxicity was not observed in the large number of acute, subchronic, and chronic studies conducted in rodents nor was it observed in two specific neurotoxicity studies conducted in dogs. However, these studies did not assess potential effects on neurotransmitters and their metabolites in the brain and other parts of the nervous system, measures of response used in current testing protocols for neurotoxicity.

Some reports on the immunotoxicity of glyphosate appear in the literature. Female CD-1 mice exposed to Roundup at concentrations up to 1.05% in drinking water for 21 d showed no change in immune function (T-lymphocyte and macrophage-dependent antibody response) when, on day 21 of the exposure period, they were inoculated with sheep erythrocytes (Blakley 1997). In an *in vitro study* on cytokine production by human peripheral blood mononuclear cells, glyphosate had only a slight effect at the greatest concentration tested (1,000 μM = 226,000 μg/L) (Nakashima et al. 2002). Results of both studies suggest that glyphosate does not affect immune response in mammals at realistic exposure concentrations. However, studies in fish suggest that that there may be some immunotoxic effects. Short exposures to Roundup (10 min at a concentration of 100,000 μg/L) in carp (*Cyprinus carpio*) and European catfish (*Silurus glanis*) caused a decrease in metabolic and phagocytic activity as well as proliferative response (Terech-Majewska et al. 2004). In contrast to these effects at large concentrations, responses on splenic antibody plaque-forming cells in the fish *Tilapia nilotica* were reported at concentrations of 1.65×10^{-2} μM (= 4.4 μg/L). As responses of the immune system are difficult to interpret in terms of survival of individuals or the population, they are not formally used in assessment of pesticides by regulatory agencies.

Toxicokinetics of glyphosate were reviewed by Williams et al. (2000). Between 15% and 36% of ingested glyphosate is absorbed through the intestinal tract and only about 2% via the skin. Excretion of unabsorbed glyphosate is via the feces, but the absorbed glyphosate is excreted via the urine with only a small amount of metabolism. Whole-body half-lives were biphasic, with an initial half-life of 6 hr and a terminal elimination half-life of 79–337 hr in rats (Williams et al. 2000). Clearance from most tissues was rapid but was cleared more slowly from the bone, possibly because of ionic binding to bone calcium (Williams et al. 2000). Glyphosate is clearly not bioaccumulated, and any absorbed dose is excreted in the urine relatively rapidly.

Cases of Human Poisoning. A number of anecdotal reports of human poisoning with glyphosate and its formulations have been published. In some cases, these are reports of a single event and an observed response.

In one, toxic pneumonitis was observed after exposure to a glyphosate formulation (Pushnoy et al. 1998). However, no information was provided to demonstrate how airborne exposure could have occurred and the results are at odds with the known inhalation toxicity of the formulation (Williams et al. 2000) and tests done on the product as used in Colombia (see following).

In another case, a man accidentally sprayed himself with an unidentified formulation of glyphosate (Barbosa et al. 2001). He developed skin lesions 6 hr after the accident but these responded to routine treatment. However, 1 mon later, the patient presented with a case of symmetrical Parkinsonism syndrome. This is an isolated case, and it is impossible to conclude anything about causality as the disease may have already been present but asymptomatic. In a similar case, a 78-yr-old woman presented with extensive chemical burns in legs and trunk caused by an accidental contact with a glyphosate formulation. These lesions disappeared, without consequences, a month later (Amerio et al. 2004).

Acute intoxication information has been documented in two case-series studies, from Taiwan, China, where glyphosate formulations were apparently used for attempted suicide (Chang et al. 1999; Lee et al. 2000). The first paper analyzed 15 intentional intoxications with glyphosate formulation and found that 68% of the patients presented esophageal, 72% gastric, and 16% duodenal injuries. Esophageal injury was the most serious injury but was minor in comparison with that caused by strong acids. Lee et al. (2000) analyzed 131 suicide attempts in southern Taiwan. The most common symptoms were sore throat and nausea; the fatality rate was 8.4%. In this study, 20.5% presented respiratory symptoms and more than half of them needed intubations. The authors propose that direct damage to the airway passage occurs and mention that the surfactant in the formulation (POEA MON 0818) may be responsible for the toxicity. In many cases, the exact doses consumed by persons attempting suicide are not known, and it is difficult to interpret these findings in the context of bystander and other accidental exposures, which are usually many orders of magnitude less. It is, however, interesting to note the low fatality rate compared to what has been reported from other pesticides such as paraquat and the organophosphorus insecticides (Krieger 2001).

It is well known that the older formulations of glyphosate that contained the surfactant POEA (MON 0818) in larger amounts were eye irritants. Goldstein et al. (2002) analyzed 815 glyphosate-related "calls" to the Pesticide Illness Surveillance Program (PISP), most of them involving eye irritation (399), skin (250), upper airway (7), and combinations of these. Of the 187 systemic cases, 22 (12%) had symptoms definitely related to exposure to formulations of glyphosate. Again, this is not surprising as the formulation of glyphosate is acidic, similar to strong vinegar, and the surfactant is an eye irritant. In other studies on eye and skin irritation reviewed in Williams et al. (2000), none of the reported exposures resulted in permanent change to the structure or function of the eye. Based on these findings, it was concluded that the potential for severe ocular effects in users of Roundup herbicides is extremely small. This observation is consistent with the minimal ocular and dermal effects observed with the formulation of glyphosate used in Colombia (see below).

Human Epidemiology Studies. A number of studies in the recent epidemiology literature have attempted to address the issue of glyphosate exposure and disease incidence in humans. Epidemiology studies on pesticides commonly suffer from two sources of error. Possibly the most important of these is the error in assigning exposures. Exposures in the studied population are never measured directly and it is common to use surrogates for exposures such as areas treated with pesticides, number of applications, and/or number of years of application. Studies have shown that these surrogates are susceptible to significant errors (Arbuckle et al. 2004). Similar conclusions have been put forward by others (Arbuckle et al. 2005; Harris et al. 2002; Solomon et al. 2005b). A second possible source of error is the fact that the populations that are studied (farmers and professional applicators) typically use many pesticides. Thus, any substance-specific responses and causality are difficult to ascertain.

Cancer Studies. The work of Hardell et al. (2002) presented a pooled analysis of two case-control studies, one on non-Hodgkin's lymphoma (NHL) (Hardell and Eriksson 1999) and another related to a hairy cell leukemia (HCL), a rare subtype of NHL. The 1999 study employed a case-control study design based on a total of 442 subjects; however, only 4 cases and 3 controls, or less than 1% of the overall study subjects, reported the use of glyphosate. The conclusions are thus based on small numbers and the confidence interval (CI) reported for exposure to glyphosate was 0.4–13, showing a lack of statistical confidence. In their pooled analysis, Hardell et al. (2002) reported a positive association with use of glyphosate [odds ratio (OR) 3.04, 95% CI of 1.08–8.52] when analyzed using univariate statistics with the highest risk for exposure during the latest decade before diagnosis. However, the OR was reduced when using multivariate statistics (OR 1.85, 95% CI of 0.55–6.20). In addition, the study was based on a small number of cases and controls (8/8) and lacked power to differentiate linkages.

De Roos et al. (2005) evaluated associations between glyphosate exposure and cancer incidence in the Agricultural Health Study (AHS), a prospective cohort study of 57,311 licensed pesticide applicators in Iowa and North Carolina. Among private and commercial applicators, 75.5% reported having ever used glyphosate, of which >97% were men. In their analysis, glyphosate exposure was defined as (a) ever personally mixed or applied products containing glyphosate, (b) cumulative lifetime days of use, and (c) intensity-weighted cumulative exposure. Glyphosate exposure was not associated with incidence of 12 common cancer types [the relative risk (RR) included 1 in all cases]; however, the RR for multiple myeloma incidence was 2.6 (95% CI of 0.7–9.4 based on 32 cases of the total of 2,088 cancers), prompting the authors to suggest that this should be followed up in future studies.

Overall, there is no strong evidence to link glyphosate exposure to increased risk of cancer. Taken with the lack of any evidence of genotoxicity or carcinogenicity of glyphosate in laboratory studies (Williams et al. 2000), it is highly unlikely that glyphosate is carcinogenic in humans.

Neurological Effects. A recent study on farmers in the Red River Valley in Minnesota (USA), reported on the link between glyphosate and attention deficit disorder and attention deficit hyperactivity disorder (ADD/ADHD) in children of farmers who applied it (Garry et al. 2002). They reported an OR of 3.6 (95% CI, 1.3–9.6); however, the study suffered from several potential sources of error. The authors noted the lack of uniform diagnostic neurobehavioral information related to ADD/ADHD and that their study identified 14 cases of ADD/ADHD among 1,532 live births, a frequency that was actually considerably lower than background rates of ADD/ADHD which had previously been reported by researchers in Canada and the U.S. Notwithstanding, while Garry et al. (2002) concluded that their study showed a tentative association between ADD/ADHD and the use of glyphosate, they also noted that other experimental evidence did not support this conclusion, including that glyphosate was not genotoxic and that little, if any, evidence of neurotoxicity has been associated with exposure to glyphosate, except in cases of intentional oral overdose. Finally, the authors expressed concern that their tentative conclusions could be explained by random chance alone and stated the need for further detailed neurodevelopmental studies to resolve these outstanding issues. Overall, there appears to be little evidence to support a link between glyphosate exposure and neurobehavioral problems in children of exposed applicators.

Reproductive Outcomes. Several papers have reported on the relationship between adverse reproductive outcomes and the use of glyphosate. In a study in Ontario, Canada, Arbuckle et al. (2001) observed a moderate increase in the risk of late abortions associated with preconception exposure to glyphosate (OR = 1.7, 95% CI, 1.0–2.9). Another study in Ontario, part of the Ontario Farm Family Health Study, reported a positive association (decrease in fecundability of 20%, ratio range = 0.51–0.80) when both spouses participated in activities where they could be exposed to pesticides. This finding was observed for 6 of 13 pesticide categories, 1 of which was glyphosate (Curtis et al. 1999). The study was based on 2,012 planned pregnancies. There was no strong or consistent pattern of associations of pesticide exposure with time to pregnancy (TTP). For exposure intervals in which only the men participated in pesticide activities or in which neither men nor women participated in pesticide activities but pesticides had been used on the farm, conditional fecundability ratios ranged from 0.75 to 1.50, with no apparent consistency among pesticide classes, chemical families, or active ingredients. Again, although this study did suggest a linkage between pesticide exposure and fecundability, there is no evidence from laboratory studies that glyphosate is a reproductive toxicant at exposures that would be expected in humans (Williams et al. 2000).

Overall, there is little epidemiological evidence to link glyphosate to any specific diseases in humans. This conclusion is supported by laboratory toxicity studies. However, responses related to reproductive outcomes such as fecundability measured through time to pregnancy offer a useful measure of possible effects that can be applied in situations such as Colombia where other health data are difficult to gather. With this in mind, a preliminary study was designed to gather human epidemiological data in several regions in Colombia. These regions were the same as those selected for the surface-water sampling (see Table 9). The design and results are summarized in the following section; a detailed report is given in a separate document (Sanin 2005).

Human Health Effects – Time to Pregnancy. A specific study was conducted to elucidate possible effects on reproductive health from exposure to glyphosate and adjuvants by assessing fertility/fecundability among women resident in different areas of the country with different pesticide use patterns. The design was cross-sectional with retrospective collection of data and is equivalent to a retrospective cohort. The study population consisted of 600 women of reproductive age in each of five different areas (see Table 9). The independent variable in the study was exposure to glyphosate for control of illicit crops, measured through use information from the region as indicated in Table 9. Possible confounders or independent predictors of reproductive variables in study the were also considered (Sanin 2005; Solomon et al. 2005a).

The distribution of pregnancies in relation to time to pregnancy (TTP) (Fig. 8) was different among the five regions. In previous work in Colombia (Idrovo et al. 2005), the percentage of pregnancies for first month was about 30%, small compared with data from developed countries. In this case, Valle del Cauca had very small initial percentage and Boyacá had larger values for the 1st and 12th months (Fig. 8). The mean for 12 mon in developed countries is 85%–90%.

In the crude analyses, longer TTP was associated with a number of factors such as region, older maternal age, ethnic group, irregular menstrual cycles, and irregular partner relationship. Previous visits to a physician for problems related with fertility, X-rays taken in the year before pregnancy (YBP), and coffee consumption in the YBP also were associated with longer TTP. Coffee consumption had a significant test for trend, but the odds ratio (OR) was not significant. Maternal overweight was associated with a longer TTP. A tendency to longer TTP was observed among those

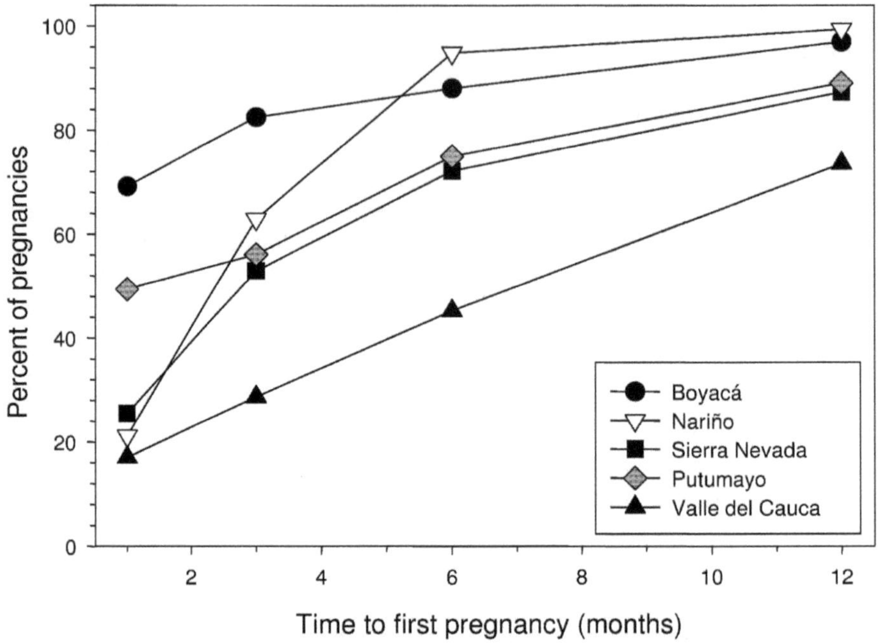

Fig. 8. Time to pregnancy (TTP) in the five study regions in Colombia.

engaged in some waged work and with higher education. Paternal unemployment or self-employment was associated with longer TTP. No other paternal data were related with the TTP.

In the final multivariate model, the main predictor of TTP was the region adjusted by irregular relationship with partner and maternal age at first pregnancy. Boyacá had the minimal risk and was the reference region; Nariño, Sierra Nevada, and Putumayo had slightly greater risk. The greatest risk was in the Valle del Cauca region. There was no association between TTP and use of herbicides in the control of illicit crops in the regions studied. The reason(s) for the increased risk for longer TTP in the Valle del Cauca region, where sugar cane is grown, is not known. In this study, the increased risk in Valle del Cauca cannot be attributed to exposure to pesticides alone because Sierra Nevada, where organic crops are grown, also showed a statistically significant difference from the reference location where pesticides are used (Boyacá). This study was designed to test hypotheses related to the use of glyphosate in eradication spraying, and the data cannot be used to identify causality associated with other risk factors. To test this question in Valle del Cauca or any other region, a new study would have to be designed and conducted. Some of the factors associated with higher TTP that were identified should be included in any future studies.

Effects of Glyphosate in Nontarget Organisms in the Environment

The mechanism of action of glyphosate is via the disruption of the shikimate metabolic pathway that leads to the synthesis of aromatic compounds in numerous microorganisms and plants. Glyphosate translocates to active growing tissues; this is particularly effective in most plants because its degradation is slow. Thus, the herbicide moves throughout the plant before symptoms are noticed. The shikimate pathway is absent from mammals (Eschenburg et al. 2003; Roberts et al. 1998, 2002). However, toxic effects of the compound on, for example, nonmammalian aquatic organisms, have been observed at large concentrations. These effects are discussed in more detail below.

A common question in conducting risk assessments in tropical regions and other nontemperate regions is the paucity of toxicity data for "tropical species." It is true that most of the test species used in toxicity testing, particularly of pesticides, are "temperate species" largely because of the location of testing laboratories that are able to conduct guideline toxicity tests under Good Laboratory Practice (GLP). Except for a few substances with defined mechanisms of action, there is no reason to believe that organisms from tropical regions are inherently more or less sensitive than organisms from temperate regions. It is well known that DDT and some of the pyrethroids become more toxic at lower temperatures (Dyer et al. 1997); however, the mechanisms here are well understood and do not apply to glyphosate. Comparison of responses of tropical and temperate organisms to a number of pesticides other than DDT has shown that there are no significant differences in sensitivity (Maltby et al. 2005). With this in mind, we used the rich data set of toxicity values that have accumulated in the literature for glyphosate and its formulations.

Effects in Nontarget Terrestrial Animals. The potential environmental effects of glyphosate and Roundup were extensively reviewed in 1999 (Giesy et al. 2000). Additional papers have appeared since then.

Soil Invertebrates. The effects of glyphosate and formulations on earthworms have been reviewed (Giesy et al. 2000) and risks were judged to be essentially negligible. A recent study on the earthworm *Eisenia fetida* reported that, although a commercial formulation of glyphosate was not directly toxic to the earthworms, it did cause effects on locomotory activity that may be detrimental to the earthworms (Verrell and Van Buskirk 2004). The formulation used in the study was Ortho Groundclear Total Vegetation Killer, which contains 5% by volume glyphosate as the isopropylamine salt (IPA). In this study, the authors applied 82 mL of a 1 : 4 solution of Groundclear to 2 L of soil in a plastic box, an amount much greater than would be applied under normal agricultural uses or in the control of illicit crops. Assuming that the boxes of soil were cubes, the area of the surface would be 12.6×12.6 cm or 159 cm^2. This being so, the application rate was equivalent to 518 kg glyphosate/ha, a totally unrealistic application rate and 100 times more than that used in the control of coca. This study was seriously flawed, and the results are not applicable to any field use of glyphosate.

Soil Microorganisms. Glyphosate and its formulations have little effect on soil microorganisms (Giesy et al. 2000). Because the symbiotic soil and root-associated microorganisms may be partially dependent on the plant for nutrients, the death or injury of the plant will affect the organisms associated with it. Similarly, death of the plants will release organic matter and nutrients into the soil, affecting soil microorganisms similarly to the application of compost or fertilizer. This response, as reported for glyphosate formulation and its effects on grass (Tenuta and Beuchamp 1995), would also occur with other herbicides and with mechanical control of plants. Effects have been demonstrated in hydroponically grown plants exposed through the watering solution; however, this route of exposure is not relevant to field conditions where glyphosate would bind strongly to soil particles and not be biologically available. Effects on symbiotic microbiota have also been demonstrated in glyphosate-tolerant plants treated at 10 times normal field application rates, but these are not relevant exposures as the studies were done in vitro and in the absence of soil (Mårtensson 1992). Some effects on metabolism of phenolic substances in symbiotic bacteria in glyphosate-tolerant soybeans have been shown; however, these changes did not alter nitrogenase activity (Hernandez et al. 1999). Microbial systems in soil are complex, and considerable variation can be expected among tests and among soil types. More-recent studies on the effects of glyphosate on microbiological activity in soils have shown an increase in microbiological activity, mainly in fungi, which are likely using the glyphosate as a source of carbon, nitrogen, and phosphorus (Araujo et al. 2003; Haney et al. 2002; Laatikainen and Heinonen-Tanski 2002). These changes in microbiological activity are not judged to be deleterious.

The effects of several fungicides and herbicides on the growth of the ectomycorrhizal fungi *Lactarius deliciosus*, strain LDF5, and *Pisolithus tinctorius*, strains 30AM, 3SR, and Mx, in pure culture have been studied. Glyphosate (formulation unspecified) at concentrations of 0, 1, 10, 100, and 1,000 mg/kg soil had no effect (Diaz et al. 2003). Some 64 strains of ectomycorhizal fungi were tested against the most common pesticides used in forestry in Finland. Glyphosate did not produce strong inhibition in any of the strains, most were unaffected, and some were stimulated by 1 mg/L Roundup Bio in agar (Laatikainen and Heinonen-Tanski 2002). Laboratory tests on four species of entomopathogenic fungi have shown that glyphosate has no effect, but a range of formulated products did have fungicidal properties, especially RoundUp Ready-To-Use (Morjan and Pedigo 2002). In fact, as fungi and bacteria have the shikimate pathway, this suggests the potential use of shikimate pathway inhibitors for the beneficial control of fungal pathogens and apicomplexan parasites, such as *Toxoplasma gondii*, *Plasmodium falciparum*, and *Cryptosporidium parvum* (Roberts et al. 1998, 2002).

Analysis of all lines of evidence for effects of glyphosate and its formulations on soil microorganisms indicates that adverse effects would be unlikely as a result of application at normal field rates. Any minor effects to communities, such as described above, would be expected to disappear rapidly (Giesy et al. 2000; World Health Organization International Program on Chemical Safety 1994). After reviewing several studies conducted in many climates, in different soils over the past 10 years, and under various cropping systems, Motavalli et al. (2004) have concluded that, so far, glyphosate and its formulations have no relevant effect on nutrient transformations by microbes. However, they point out that this topic needs further study, as not every situation has been adequately researched. Further, because of lack of bioavailability on soils, adverse effects on beneficial soil fungi and bacteria are unlikely to occur under field conditions. Glyphosate binds strongly to soil particles and would not be available for uptake by microorganisms, many of which are actually inside the plant tissues. The fact that seeds will readily germinate in soils soon after treatment with glyphosate and that nitrogen-fixing Roundup Ready soybeans grow and develop high yields despite treatment with glyphosate demonstrates the practical insignificance of these effects under actual use conditions.

Terrestrial Invertebrates. As glyphosate is a nonselective herbicide, it will cause habitat alteration, which also results from a number of human activities in the production of food and fiber. Most important is the clearing of land for agricultural production. Whether this is through slash-and-burn processes as used in the initial preparation of coca and poppy fields or the application of herbicides such as glyphosate and paraquat, also used in coca production, the effects on nontarget species are the same. Use of cultural or mechanical controls, or herbicides, to remove plants will have effects on organisms that normally use these plants for food or shelter.

After applying glyphosate formulation at double the recommended application rates, no effects were observed in microarthropods in soil (Gomez and Sagardoy 1985). Because weed species compositions and densities are directly affected by glyphosate, indirect effects are more likely to occur. Jackson and Pitre (2004b) found that populations of adult *Cerotoma trifurcata*, adult *Spissistilus festinus*, larvae of *Plathypena scabra*, and the caterpillar of *Anticarsia gemmatalis* were unaffected by glyphosate formulation, but populations of adult *Geocoris punctipes*, a homopoteran insect predator, were decreased. This effect was caused by reduced weed densities after herbicide treatment. Populations of green cloverworm (*Hypena scabra*) were evaluated on soybean glyphosate-resistant varieties, with and without exposure to glyphosate [glyphosate acid Roundup Ultra at 2.48 kg equivalent per hectare (AE/ha)], and no differences among treatments were detected on developmental time and survivorship (Morjan and Pedigo 2002). Weed management systems, more than glyphosate, that allowed more weeds to grow generally had higher insect population densities (Buckelew et al. 2000).

Effects of glyphosate and associated cultural practices can affect arthropods indirectly. In studies conducted in the U.K., indirect effects of glyphosate were observed in the spider *Lepthyphantes tenuis*, caused by habitat alteration and related to death of plants and decreasing height of vegetation. Applications of glyphosate had only a within-season indirect habitat effect on *L. tenuis* as field margins sampled 16 mon after an application of 360 g glyphosate AE/ha showed no detrimental effects (Bell et al. 2002; Haughton et al. 2001). Tests of the fecundity and mortality of *Geocoris punctipes* (Say), exposed to glyphosate as Roundup on soybean found no effects over 10-d posttreatment. Exposure of *G. punctipes* eggs to Roundup Ultra spray had no effect on egg hatch (Jackson and Pitre 2004a). Some reductions in numbers of this species 3 wk after treatment probably reflect weed removal (Jackson and Pitre 2004b).

Similarly, studies on populations of leaf litter invertebrates in areas of Australia where Roundup Biactive was sprayed at 1–1.4 kg/ha for the control of an invasive weed showed no significant effects 4 mon after spraying (Lindsay and French 2004). Variability in treated and untreated areas was large and suggested that the nature of the vegetative community and its structure and the postspray weather may also be important. In agriculture, these effects are part of the risk assessment related to integrated pest management (IPM), and potential effects on beneficial organisms are weighed in the risk–benefit equation. In conclusion, there is little evidence of any direct effect of glyphosate on insects in the field or in natural environments.

Terrestrial Vertebrates. Technical glyphosate, formulated glyphosate (above), and glyphosate mixed with Cosmo-Flux (see below) are not acutely toxic to mammals via several routes of exposures. Although wild mammals have not been specifically tested with the mixture as used in Colombia, data from these laboratory studies suggest that they would be insensitive and not directly affected by a direct overspray.

Birds are not susceptible to glyphosate. In studies on bobwhite quail (*Colinus virginianus*) and mallard duck (*Anas platyrhynchos*), acute oral LD_{50} values of >4,640 and >4,640 mg/kg bw have been reported (USEPA 2001). Again, direct effects of formulated glyphosate or glyphosate plus Cosmo-Flux are judged to be very unlikely.

Indirect effects on terrestrial wildlife have been reported with the use of glyphosate in agriculture and forestry. Alteration of habitat is more of an issue in semiwild areas such as forests where herbicides may be used to control competing vegetation and allow conifers to grow and mature more rapidly. In these cases, short-term effects on birds and other wildlife do occur; however, these populations usually recover in 2–3 yr (Kimball and Hunter 1990; Santillo et al. 1989a,b), and even the vegetation recovers in less than 10 yr (BC Ministry of Forests 2000; Boateng et al. 2000). Normally, in these uses, the actual areas treated are relatively small and are surrounded by or adjacent to untreated areas that can act as refugia or sites for repopulation by animals that have moved away because of the changes in habitat. As new vegetation develops to replace that controlled by the herbicide, the habitat will again become usable by these animals (Giesy et al. 2000; World Health Organization International Program on Chemical Safety 1994).

Glyphosate is widely used for vegetation management, including the restoration of native plant communities where exotic or invasive species are controlled (Hartman and McCarthy 2004). The use of glyphosate for "conifer release" from competition has minimal effects on wildlife and can be used to enhance biodiversity if used for spot and patch treatments (Sullivan and Sullivan 2003). A review of management of northern U.S. forests, including the use glyphosate, indicated no adverse ecological effects (Lautenschlager and Sullivan 2002). However, the impacts of vegetation removal by manual clearance and glyphosate application in conifer plantations had effects on bird communities in British Columbia, mediated by the removal of deciduous plants. Where the herbicide was used, number of bird species declined, total number of individuals increased, and common species dominated. Populations of residents, short-distance migrants, ground gleaners, and conifer nesters increased significantly after herbicide treatment. Deciduous nesters and foliage gleaners increased in abundance (nonsignificantly) in control and manually thinned areas. Warbling vireos (*Vireo gilvus*), which are deciduous specialists, declined in treated areas and may be particularly susceptible to the indirect effects of glyphosate plant removal (Easton and Martin 1998; 2002).

Nevertheless, control of *Cirsium arvense* (Canada thistle) using wick application of glyphosate in wildfowl areas can enhance plant diversity that benefits water birds (Krueger-Mangold et al. 2002). However, the broad-spectrum activity of glyphosate means that accidental overspray of rare nontarget plant species during control of invasive plants will cause damage (Matarczyk et al. 2002).

Beneficial Insects. Glyphosate is not considered toxic to honeybees, with a reported LD_{50} of >100 µg/bee (USEPA 2001), however, the formulation, with the adjuvant Cosmo-Flux, as used in Colombia may have different toxicity because of the added surfactants. To test this hypothesis, toxicity testing of a mixture of a commercial formulation of glyphosate and the surfactant Cosmo-Flux 411F was conducted to determine the acute contact toxicity to honey bees (*Apis mellifera* L.) (Stantec 2005a), following standard test guidelines (OECD 1998a; USEPA 1996a). The mixture of glyphosate and Cosmo-Flux 411F was not toxic via acute contact exposure to honeybees (i.e., did not cause mortality or stress effects in bees within 48-hr of treatment) at concentrations equal to or less than 63.9 μg AE/bee. These results are similar to those for glyphosate and formulations from the USEPA ECOTOX data base (USEPA 2001) and show that the formulated product as used in Colombia is not hazardous to bees or, by extrapolation, to other beneficial insects.

Effects in Aquatic Animals. Several extensive reviews of the effects of glyphosate on aquatic organisms have concluded that glyphosate presents an essentially negligible risk to aquatic organisms (Giesy et al. 2000; Solomon and Thompson 2003; World Health Organization International Program on Chemical Safety 1994). Several publications reported on the effects of glyphosate and several of its formulations in frogs. The acute toxicity of technical-grade glyphosate acid, glyphosate isopropylamine, and three glyphosate formulations to Australian frog species was measured by Mann and Bidwell (1999). Acute toxicity was observed for adults of one species and tadpoles of four species of southwestern Australian frogs in 48-hr static/renewal tests. The 48-hr LC_{50} values for Roundup herbicide (MON 2139) tested against tadpoles of *Crinia insignifera, Heleioporus eyrei, Limnodynastes dorsalis*, and *Litoria moorei* ranged between 8,100 and 32,200 μg/L (2,900 and 11,600 μg AE/L, whereas the 48-hr LC_{50} for Roundup herbicide tested against adult and newly metamorphosed *C. insignifera* ranged from 137,000 to 144,000 μg/L (49,400–51,800 μg AE/L). These values were different, depending on the type of dilution water (lake or tap water). For the purposes of this risk assessment, the most sensitive stage was used.

Touchdown herbicide (4 LC-E) tested against tadpoles of *C. insignifera, H. eyrei, L. dorsalis*, and *L. moorei* was slightly less toxic than Roundup with 48-hr LC_{50} values between 27,300 and 48,700 μg/L (9,000 and 16,100 μg AE/L, respectively). Roundup Biactive (MON 77920) was practically nontoxic to tadpoles of the same four species, producing 48-hr LC_{50} of 911,000 μg/L (328,000 μg AE/L) for *L. moorei* and >1,000,000 μg/L (>360,000 μg AE/L) for *C. insignifera, H. eyrei*, and *L. dorsalis*. Technical glyphosate isopropylamine salt was practically nontoxic, producing no mortality among tadpoles of any of the four species over 48-hr, at concentrations between 503,000 and 684,000 μg/L (343,000 and 466,000 μg AE/L). The toxicity of technical-grade glyphosate acid (48-hr LC_{50}, 81,200–121,000 μg AE/L) is likely to be caused by acid intolerance. Slight differences in species sensitivity were evident, with *L. moorei* tadpoles showing greater sensitivity than those of the other four species. Adult and newly emergent metamorphs were less sensitive than tadpoles.

A series of studies on frogs was conducted with several formulations of glyphosate in relation to its use in forestry in Canada (Chen et al. 2004; Edginton et al. 2004; Thompson et al. 2004; Wojtaszek et al. 2004). Using a formulation of glyphosate (Vision) containing glyphosate and ethoxylated tallowamine surfactant (POEA), LC_{50} values as low as 880 μg AE/L were reported for tadpoles of *Xenopus laevis, Bufo americanus, Rana clamitans*, and *Rana pipiens* (Edginton et al. 2004). Embryo stages were less sensitive than older larvae, and toxicity was affected by the pH of the exposure medium, although not in a consistent manner. For the purposes of this assessment, values obtained at the most sensitive pH and for the most sensitive stage were used.

In a related study on the toxicity of the Vision formulation of glyphosate to the zooplankton organism *Simocephalus vetulus* and tadpoles (Gosner stage 25) of *Rana pipiens,* interactions between pH and food availability were reported (Chen et al. 2004). Both high pH (7.5 vs. 6.5) and food deprivation increased the toxicity of this formulation. As only two concentrations were tested (750 and 1,500 μg AE/L), LC_{50} values could not be determined.

Field studies conducted on larvae of *R. clamitans* and *R. pipiens* with Vision showed that, in the presence of natural factors such as sediment and environmentally relevant pH, toxicity of the formulation was reduced compared with laboratory observations (Wojtaszek et al. 2004). The authors reported 96-hr LC_{50} values ranging from 2,700 to 11,500 μg AE/L (Wojtaszek et al. 2004). Although they used a formulation of glyphosate containing the more-toxic surfactant POEA, the results confirm that, in the presence of sediments, reduction in the bioavailability of glyphosate (and formulants) occurs, further reducing risks, a conclusion reached for this forestry use (Thompson et al. 2004) but which is equally relevant to the use of glyphosate in Colombia. These observations are consistent with the rapid dissipation of both glyphosate and the POEA surfactant in the presence of sediments (Tsui and Chu 2004; Wang et al. 2005).

Toxicity of a number of glyphosate formulations to frogs (*R. clamitans, R. pipiens, Rana sylvatica*, and *Bufo americanus*) was reported (Howe et al. 2004); these included Roundup Original, glyphosate technical, the POEA surfactant used in some glyphosate-based herbicides, and five newer formulations of glyphosate. As expected, the most toxic of the materials was the POEA surfactant, followed by Roundup Original, Roundup Transorb, and Glyfos AU. No significant acute toxicity was observed with glyphosate technical material [96-hr LC_{50} > 17,900 μg/L(AE)]. LC_{50} values for Roundup Original in *R. clamitans*, *R. pipiens*, and *R. sylvatica* were 2,200, 2,900, and 5,100 μg AE/L, respectively; these values were used in this risk assessment. Several other formulations of glyphosate were also tested in *R. clamitans* (Roundup Biactive, Touchdown, and Glyfos BIO) and were essentially nontoxic with LC_{50} values >57,000 μg AE/L.

In a study on *Rana cascadae,* a 48-hr LC_{50} for Roundup (52% IPA) of 3,200 μg AI/L (2,336 μg AE/L) was reported using static exposures in glass tanks (Cauble and Wagner 2005). In a chronic exposure study with the same formulation over a 43-d period in glass tanks without sediment using a 7-d static renewal exposure with nominal concentrations of 730 and 1,460 μg AE/L, the authors found a number of effects such as decreasing time to death, increased mortality (8.6% and 51% at 730 and 1,460 μg AE/L, respectively), increased time to metamorphosis, and decreased weight of metamorphs. The relevance of these observations to exposures under field conditions must be considered. Under field conditions, glyphosate and its surfactants have been shown to bind strongly to sediments, which rapidly reduces concentrations, resulting in reduced toxicity. Thus, under use conditions in the field, exposures longer than the 48–96-hr used in acute tests are extremely unlikely. As no sediments were present in the glass tanks of this study, the chronic exposures used are not representative of what occurs in the field and are not applicable to the risk assessment for amphibians.

In a study of several commercial pesticide formulations in leopard frogs (*Rana pipiens*), green frogs (*R. clamitans*), bullfrogs (*Rana catesbeiana*), the American toad (*B. americanus*), and gray tree frogs (*Hyla versicolor*), effects of Roundup and interactions with other pesticides were reported (Relyea 2004). The Roundup used in this study contained the more-toxic POEA surfactant. Survival and growth over a 16-d period were not significantly affected by the glyphosate formulation at 1,000 µg AE/L but some species were affected at 2,000 µg AE/L. Some interactions were observed between glyphosate and other pesticides such as the insecticides diazinon, carbaryl, and malathion. One paper reported that a glyphosate formulation containing POEA was highly toxic to tadpoles of several species of frogs exposed under realistic conditions in small (1,000-L) field microcosms (Relyea 2005a). The tadpoles (wood frog, *Rana sylvatica*; leopard frog, *Rana pipiens*; American toad, *Bufo americanus*; gray tree frog, *Hyla versicolor*; and the spring peeper, *Pseudacris crucifer*) were exposed to 3,800 µg AE/L commercial glyphosate (unspecified) applied directly to the surface of the water. Application rate was equivalent to 16 kg AE/ha, an unrealistic value. At this concentration, glyphosate formulated with POEA would be expected to be lethal to tadpoles. The discussion in the paper that suggests that use of glyphosate may have adverse effects on frogs more generally is thus based on a flawed study design and is not supported by other data, as already discussed. In a laboratory study (Relyea 2005b) in which juvenile terrestrial stages of three different species (*R. sylvatica, B woodhousii folweri, and H. versicolor*) were exposed to direct applications of formulated glyphosate at 1.6 mg active ingredient (AI)/m^2 (1.2 mg AE/m^2) in plastic tubs, 79% mortality was observed after only 24-hr. The volume of formulation (6.9 mL) used to spray the tubs and the concentration of glyphosate (1.9%, IPA assumed) suggest that actual exposures were much greater than stated (91 mg AE/tub). Clearly, there were errors in the description of the methods, and the results of the study are uninterpretable.

Effects on other nontarget aquatic organisms have also been reported. In studies on the toxicity of glyphosate to several aquatic algae and zooplankton, Tsui and Chu (2003) showed that technical glyphosate was considerably less toxic than Roundup, which is formulated with the POEA surfactant. LC_{50} and EC_{50} values for technical glyphosate ranged from 5,890 to 415,000 µg AE/L. In tests conducted in the presence of sediment (Tsui and Chu 2004), they showed that biological availability of glyphosate was significantly reduced by binding to sediment. The reduction in concentration in pore water resulting from the sediments was proportional to the amount of organic carbon in the sediments.

Tests on the fish *Oreochromis niloticus* (Nile tilapia) exposed for 3 mon to sublethal concentrations (5,000 and 15,000 μg/L) of glyphosate as Roundup caused significant damage to gill, liver, and kidney tissue. The structural damage could be correlated to the significant increase ($P \leq 0.05$) in aspartate aminotransferase, alanine aminotransferase, and alkaline phosphatase activities in the second and third months of exposure. The results indicated that long-term exposure to Roundup at large, although at sublethal concentrations, had caused histopathological and biochemical alterations of the fish (Jiraungkoorskul et al. 2003). Because technical glyphosate was not tested, the contribution of surfactants to this response cannot be judged.

In studies on the freshwater mussel *Utterbackia imbecillis*, a commercial formulation of Roundup was reported to have little toxicity (24-hr LC_{50} of 18,300 μg/L and a NOEC of 10,040 μg/L–7,442 μg AE/L) to larval mussels (Conners and Black 2004). In studies on genotoxicity in these mussels, there was no significant difference in response between the control and mussel larvae treated at one-fourth the NOEC, ≈2,500 μg/L (1,850 μg AE/L).

Response of total free amino acid profiles of snails to glyphosate exposures has been studied (Tate et al. 2000), showing that exposure of the aquatic snails (*Pseudosuccinae columella*) to technical glyphosate at nominal concentrations of 1,000–10,000 μg/L led to increased egg laying and increased amino acid concentrations in tissues. Technical glyphosate was not particularly toxic with a 24-hr LC_{50} of 98,900 μg/L (72,200 μg AE/L). The effect on egg laying and amino acid concentrations was stimulative rather than adverse, but the authors speculate that it could lead to increases in incidence of diseases for which the snails are intermediate hosts. Increases in parasites may affect organisms in the environment. Similar stimulation was observed in the rotifer *Brachionus calyciflorus* in which growth rates and sexual and asexual reproduction were stimulated in the presence of glyphosate (formulation unknown) at ≥4,000 μg/L (growth) and ≥2,000 μg/L for reproduction and resting egg production (Xi and Feng 2004). Again, although stimulatory and not "adverse", the authors point out that increases in one species may affect other species indirectly.

In a study on grazing of the alga *Scenedesmus* spp. by the aquatic crustacean *Daphnia pulex*, technical glyphosate was shown to have no adverse effect, although it appeared to stimulate algae growth (Bengtsson et al. 2004). This stimulation was suggested to be caused by release of nitrogen and phosphorus from glyphosate metabolism by *Daphnia*. Similar stimulation was also seen in the effects of glyphosate (Rodeo, glyphosate IPA without surfactants) on the primary productivity of a natural phytoplankton algal assemblage dominated by species of diatoms and a dinoflagelate (Schaffer and Sebetich 2004). A 60% increase in productivity as measured by assimilation of $^{14}CO_2$ was observed at concentrations of 125, 1,250, and 12,500 μg/L, with no apparent concentration response. The authors speculate that the increase was caused by the release of nitrogen and phosphorus from the breakdown of glyphosate.

Effects of Glyphosate on Terrestrial Plants

There are differences in glyphosate uptake between different coca species and between young and mature plants of *Erythroxylum coca* and *E. novogranatense* (Ferreira and Reddy 2000). Absorption through the leaf is greater in young plants of both species and greater in *E. novogranatense.* Earlier studies showed that control of regrowth was better in *E. novogranatense* for equivalent rate of glyphosate (Ferreira et al. 1997). This study also indicated that defoliation of *E. coca* 24-hr before application resulted in no significant effect of glyphosate (applied up to 6.7 kg AI/ha) on regrowth. This result confirms that, as for other plants, uptake via the leaves is the major route of penetration into the plant.

A study on the control of the perennial weed pepperweed (*Lepidium latifolium*) has shown better control with glyphosate following mowing. The mechanism is via the better movement of glyphosate to roots from leaves lower in the canopy. Following mowing, the leaf distribution and the spray deposition are closer to the ground, giving better basipetal translocation to roots and better subsequent control (Renz and DiTomaso 2004). In forestry situations with an aerial application, spray deposition is typically much greater higher in the canopy (Thompson et al. 1997). Studies of glyphosate efficacy on annual weeds indicated that application during the day (0900 and 1800) gives best control (Martinson et al. 2002; Miller et al. 2003).

Resistance to glyphosate is known for an increasing number of species, including *Conyza canadensis* (Mueller et al. 2003), Illinois waterhemp (*Amaranthus rudis* and *A. tuberculatus*) (Patzoldt et al. 2002), *Eleusine indica* (Baerson et al. 2002), *Lolium multiflorum* (Perez and Kogan 2003), and *Lolium rigidum* (Neve et al. 2003a,b). Rates of evolution of resistance in the latter species are dependent on herbicide use patterns as part of crop production.

Nontarget impacts of glyphosate on seed germination and growth characteristics of the F_1 generation of treated wild plant species have been reported. Blackburn and Boutin (2003) noted effects on 7 of 11 species tested with 1%, 10%, or 100% of a 0.89 kg AI/ha label rate of glyphosate formulated as Roundup solution sprayed near seed maturity. Effects of glyphosate drift on rice seed germination were reported by Ellis et al. (2003), and May et al. (2003) noted reduced seed production in alfalfa in the year following applications 1.760 kg AI/ha for *Cirsium arvense* control. Nevertheless, applications at 0.420 kg AE/ha on susceptible soybean had adverse effects on sprayed plants but not on progeny (Norsworthy 2004). Subtle adverse effects of glyphosate on pollen viability and seed set in glyphosate-resistant cotton were noted by Pline et al. (2003). Pollen viability of glyphosate-resistant corn was also significantly reduced by glyphosate applied at 1.12 kg AI/ha, but yield and seed set were not significantly affected (Thomas et al. 2004). These data indicate that drift might cause subtle ecological changes to plant communities associated with changes in plant recruitment. However, this would be significant only for communities largely made up of monocarpic plant species (those that flower once and die, especially annuals) dependent on seeds for recruitment.

B. Glyphosate and Formulants

There are a number of formulations of glyphosate on the market that may contain a number of surfactants and other formulants (Giesy et al. 2000; Solomon and Thompson 2003; Williams et al. 2000). Normally, this would not be an issue in the risk assessment of a pesticide; however, in the case of glyphosate IPA, the active ingredient has little toxicity to nontarget organisms, thus making the surfactant toxicity more important for risk assessment. For example, tests on Ca^{2+}-activated ATPase and cholinesterase (ChE) activities in the nervous system of the slug *Phyllocaulis soleiformis* showed no effects of pure glyphosate IPA. An effect noted with the formulation Gliz 480CS was caused by nonglyphosate components of the formulation (da Silva et al. 2003). Technical-grade glyphosate at concentrations of 52 mM (870 mg/L) did not affect the protozoan *Tetrahymena thermophila* or the parasite *Ichthyophthirius multifiliis*. However, the commercial formulation Glyphosate was up to 100 times more toxic, reflecting data for fish species and other aquatic invertebrates and caused by surfactants in the formulation (Everett and Dickerson 2003).

Because the spray solution as used in the control of coca and poppy in Colombia contains surfactants as part of the glyphosate formulation as well as additional surfactants (Cosmo-Flux) added to the spray mix, the toxicity of the formulants and the adjuvants may interact to change the toxicity of the mixture. For this reason, standardized toxicity tests for mammals and environmental nontarget organisms were conducted with the spray mixture itself, as discussed next.

Effects of Formulated Glyphosate and Cosmo-Flux on Mammals

Two series of mammalian toxicity tests on the formulation of glyphosate and Cosmo-Flux were conducted. One set was conducted in the U.S. (Springborn 2003a–g) under good laboratory practices (GLP) and using quality control assurance as appropriate for regulatory decision making. The other studies (Immunopharmos 2002a–j) were conducted in Colombia, also in compliance with GLP and according to USEPA guidelines. These studies were reviewed in detail (Solomon et al. 2005a), and a number of conclusions were drawn for the mixture glyphosate and Cosmo-Flux sprayed on either poppy or coca.

- The acute oral and dermal LD_{50} was >5,000 mg/kg bw in the rat. Therefore, this formulation is considered as practically nontoxic orally.
- The acute inhalation LC_{50} was >2.60 mg/L in the rat. In one study, the rats showed breathing abnormalities after exposures at 2.6 mg/L for 4-hr. This value for the test substance is considered as potentially harmful for durations of exposure of the order of 4-hr. In two other studies, the mixture was shown to be not harmful at exposures up to 20 mg/L for 4-hr. Exposures via inhalation in these studies were via small droplets. Exposures via inhalation under field conditions will be smaller as the droplets are larger and less easily inhaled.
- The formulation is considered to be a slight and moderate irritant to the skin and eyes of the rabbit. The calculated primary irritation index for the test article was 0.25.

Based on these observations, the hazards to humans via application or bystander exposures are considered small and are limited to slight to moderate skin and eye irritation. These responses will be reduced if the affected areas are rinsed shortly after exposure to remove contamination. It was also concluded that the addition of the adjuvant Cosmo-Flux to glyphosate did not change the toxicological properties of the glyphosate formulation to mammals.

Effects of Formulated Glyphosate and Cosmo-Flux on Nontarget Aquatic Organisms

A base set of toxicity data is required for all pesticide registrations. For freshwater environments, the set normally makes use of a cold-water fish, such as rainbow trout fingerlings (*Onchorynchus mykiss*); a warm-water fish, such as fathead minnows (*Pimephales promelas*); an invertebrate such as the water flea (*Daphnia magna*); and an alga, such as *Selanastrum capricornutum*. Recognized guidelines were used for the tests (OECD 1984a,b, 1992; USEPA 1996b), which were conducted under the principles of GLP (OECD 1998b). These are standard test organisms, have been used for testing glyphosate itself and several other formulations, and thus are useful for comparison purposes. To reduce the requirement for animals in the testing, one combination of glyphosate and Cosmo-Flux, the combination for poppy (see Table 2), was selected. The results of these tests are summarized in Table 11.

C. Effects in the Field

Duration of Effects

In tropical forest situations, similar to some locations in coca eradication programs, there are limited data on vegetation recovery following glyphosate application. Nevertheless, there are a number of studies of successional patterns following land clearance and for tree gaps. Forest clearance has

Table 11. Toxicity values obtained from toxicity tests conducted on a mixture of glyphosate and Cosmo-Flux.

Test species	Common name	96-hr LC_{50}/EC_{50} in μg/L (as glyphosate AE)	Reference
Selenastrum	Algae, based on cell numbers, area under the growth curve, and growth rate	2,278–5,727[a]	Stantec (2005e)
Daphnia magna	Water flea, mortality	4,240 (3,230–5,720)[b]	Stantec (2005b)
Onchorynchus mykiss	Rainbow trout, mortality	1,847 (1,407–2,425)[b]	Stantec (2005d)
Pimephales promelas	Fathead minnow, mortality	4,600 (1,805–11,700)[b]	Stantec (2005c)

[a]Greatest and smallest effect measures in the study.
[b]LC_{50}/EC_{50} and 95% confidence Interval.

been a historical feature of the development of agriculture across the globe (Boahene 1998; Matlack 1997). In Central America, agricultural intensification and forest clearance in Mayan and other cultures has been determined from the pollen record (Clement and Horn 2001; Curtis et al. 1998; Goman and Byrne 1998). Patterns of successional change (recovery) in Neotropical forests have been reviewed by Gauriguata and Ostertag (2001). The authors noted that "the regenerative power of Neotropical forest vegetation is high, if propagule sources are close by and land use intensity before abandonment has not been severe." However, they also caution that recovery is heavily dependent on interactions between site-specific factors and land use, "which make it extremely difficult to predict successional trajectories in anthropogenic settings."

In relation to the eradication program, patterns of vegetation recovery will be dependent on size of plot, location of plot in relation to surrounding vegetation types, and local anthropogenic management, i.e., subsequent cultivation activities.

A study of tree regeneration in dry and humid selectively logged Bolivian tropical forests indicated that tree release with glyphosate in logging gaps had no significant impact on target tree species growth (Pariona et al. 2003). Although glyphosate controlled vegetation for a limited period, there were problems with the recruitment of commercial trees in logging gaps, suggesting a silvicultural need for site preparation treatments and more judicious seed tree retention.

Glyphosate has been widely used for controlling deciduous understory vegetation in managed northern forests, so-called conifer-release treatments (Lautenschlager and Sullivan 2002). Recovery of the deciduous herb and shrub layers occurs over 2–3 yr in general and the tree layer over 10 yr. Often, total structural diversity is unaffected by glyphosate treatments after 1 yr.

Forest Clearance and Soils. The impacts of forest clearance on soil fertility are generally well understood. Typically, tropical forest soils are fragile, being nutrient poor and subject to leaching. Tree clearance can quickly result in loss of nutrients, change in pH, and therefore change in element availability to plants (McAlister et al. 1998). Such conditions often allow only shifting cultivation under subsistence production, so-called slash-and-burn agriculture. Studies in Jamaican forests have shown that cultivations result in large amounts of soil erosion compared with secondary forest. An agroforestry treatment with *Calliandra calothyrsus* contour hedges reduced erosion and increased rainfall infiltration within the hedges (McDonald et al. 2002). As coca is a shrub, typically grown in rows, it might be argued that soil and water changes associated with forest clearance may be less than for annual crops such as maize, but clearly both have significant adverse effects on primary forest sites.

Although vegetation recovery may be rapid in eastern North America, research has led to the surprising conclusion that 19th-century agricultural practices decreased forest floor nutrient content and C:N and C:P ratios and increased nitrifier populations and net nitrate production, for approximately a century after abandonment (Compton and Boone 2000). The level of agricultural intensity, in terms of cultivation and fertilizer use, may have significant long-term impact on soils.

Effects on Associated Fauna. In an area of highly disturbed tropical dry forest in Cordoba Department, northern Colombia, small mammals were censused by live-trapping, running from secondary growth forest into agricultural areas (Adler et al. 1997). The results suggested the disturbed habitat supports a small mammal fauna of low diversity. However, several of the species appear to have benefited from forest clearance and agricultural activities and may occasionally reach extremely high numbers, although populations were not stable. A similar effect on reduced diversity of termites with increasing disturbance has been shown in dry forest in Uganda (Okwakol 2000). Changes in bird populations of a eucalypt forest in Australia following clear-felling indicate that full recovery may take up to 70 yr (Williams et al. 2001).

Although some species are adapted to disturbed conditions and can utilize agricultural land and secondary forest, there are many species associated with primary forest only, for example, the great argus pheasant in Indonesian tropical forests (Nijman 1998). With much of Colombia associated with extremely high biodiversity, there are very many endemic plant and animal species associated with national parks and in all likelihood with areas where coca and poppy are grown.

Studies on the impacts of vegetation change caused by glyphosate use on associated fauna in northern environments are available for some species. For example, following the application of glyphosate in clear-cut forest areas in Maine (USA), the use by moose (*Alces alces*) of treated and untreated areas was compared 1–2 yr and 7–11 yr postapplication (Eschenburg et al. 2003; Eschholz et al. 1996). At 1 and 2 yr posttreatment, tracks of foraging moose were 57% and 75% less abundant on treated than untreated clear-cuts ($P = 0.013$). However, at 7–11 yr posttreatment, tracks of foraging moose ($P = 0.05$) and moose beds ($P = 0.06$) were greater on treated than untreated clear-cuts. Less foraging activity at 1–2 yr posttreatment appeared to be the result of reduced browse availability, because conifer cover for bedding was similar on treated and untreated clear-cuts. The authors hypothesized that the greater counts of tracks of foraging moose on older treated clear-cuts were the result of increased foraging activity on sites with more abundant conifer cover (Eschholz et al. 1996; Raymond et al. 1996); i.e., tree cover had returned sufficiently after 10 yr. Studies of responses of small mammals to vegetation control with glyphosate in similar environments (Sullivan et al. 1998) indicated that vegetation recovery 2–3 yr after treatment was sufficient to return population dynamics to expected ranges.

Spot applications of glyphosate to reduce invasive ground flora in forests can have the beneficial effect of opening up the ground layer and encouraging spring ephemeral species to establish larger populations. Carlson and Gorchov (2004) reported this effect when controlling *Alliaria peteolata*, an invasive biennial plant. The impact of glyphosate on the target species was only for a single season.

Recovery from Effects. Glyphosate, as a well-translocated herbicide, affects most plant species if sufficient herbicide can penetrate plant tissues, particularly leaves. Effects typically result in plant death over 2–3 wk, although species with extensive storage organs, e.g., long rhizomes, large size, or particularly impenetrable leaf surfaces, may survive. A small dose of glyphosate can result in growth abnormalities in plants, most typically localized accelerated branching. If the dose of herbicide is insufficient to cause death, it has been proposed that plant fitness may also be reduced, such that if there is competition with other plants, death may result indirectly, though there is little published evidence for this.

The effect of glyphosate is limited to the plants that receive spray at the time of application, as the herbicide is rapidly adsorbed onto soil and root uptake does not occur. The broad spectrum of plant species controlled and the pattern of foliar uptake, together with the safety of the compound, have led to widespread use of the herbicide for total vegetation control, in preharvest weed control in annual crops, and for the control of perennial plants.

Recovery of treated areas is dependent on the initial level of control, the quantities of material (and the methods used) for plant regeneration and the environmental conditions of the site. Plants have a variety of adaptations for regenerating, with some life forms showing a range of methods, while others have only a single strategy. Monocarpic species, typically annuals, have seeds for recruitment of the next generation. Polycarpic species may also produce seeds, but many also have a variety of vegetative means of regenerating, such as rhizomes, bulbs, corms, and runners. Patterns of secondary succession, the resultant plant communities over time, reflect the plant–environment interactions and the opportunities for regeneration provided by the local species pool. Seeds in the soil or those that can reach a site from the surroundings, together with vegetative fragments, will establish initially. Continued agricultural operations, such as cutting or soil disturbance, will have a major influence on the species that survive. In most situations, vegetation recovery is rapid, with ruderal and pioneer plant species establishing within weeks of application.

In tropical forests, similar to some of the locations of the coca eradication programs, there are limited published data on vegetation recovery following glyphosate application. Nevertheless, there are a number of studies of successional patterns following land clearance and for tree gaps. Secondary succession (forest recovery) has become more common in some forest areas, for example, in Puerto Rico (Chinea 2002). Forest recovery is generally fairly rapid, but recovery of the full complement of forest species can take many years (>30 yr), and the effects of bulldozing for initial clearance can reduce diversity of native species and enhance establishment of nonnative species. Comparisons of different aged plots (2–40 yr) in the Bolivian Amazon forests have contributed to the knowledge of secondary succession (Pena-Claros 2003). Not surprisingly, it takes longer for the forest canopy to achieve similar diversity to mature forest compared with the understory and subcanopy communities.

In relation to the eradication program, patterns of vegetation recovery will be dependent on size of plot, location of plot in relation to surrounding vegetation types, and local anthropogenic management, i.e., subsequent cultivation activities. Nevertheless, it should be noted that naturally occurring tree gaps (20–460 m^2) are an important component of overall forest diversity, providing opportunities for understory and subcanopy species and regeneration of canopy species in the modified light climate (Martins and Rodrigues 2002; Martins et al. 2004). In Brazilian varzea (white-water) forests, natural patterns of succession are affected by both light and local flooding (Wittmann et al. 2004). The patch scale of eradication applications of glyphosate may or may not be at the scale of natural forest gap dynamics; this deserves further study.

In the high Andes alpine paramo habitats, patterns of succession were described by Sarmiento et al. (2003). Following cultivation, usually for potato, patterns of secondary succession were such that, after 12 yr, species diversity of the undisturbed paramo had still not been attained. The characteristic paramo life forms, sclerophilous shrubs (e.g., *Baccharis prunifolia*, *Hypericum laricifolium*) and giant rosettes (e.g., *Espeletia schultzii*), appear very early and gradually increase in abundance during succession (Sarmiento et al. 2003).

In situations of agricultural expansion over large areas in Europe and North America, there is evidence that, where the proportion of remaining ancient habitat is small, subsequent forest recovery on abandoned agricultural land can be extended over long time periods (Vellend 2003). It is unlikely that habitat fragmentation and intensity of agriculture will combine to provide such a scenario in the coca eradication areas.

Effects on the successional patterns of vegetation in northern temperate and boreal forest situations are that woody and herbaceous species are most reduced by glyphosate (Bell et al. 1997). In a study in British Columbia, species richness, diversity, and turnover of the herb, shrub, and tree layers were not significantly ($P > 0.10$) different between mechanical and glyphosate spray cut stump treatments and a control. Similarly, the structural diversity of herb, shrub, and tree layers were also not significantly different ($P > 0.10$) between treatments and control. By opening the canopy and decreasing the dominance of the deciduous tree layer, both manual and cut-stump treatments showed greater total structural diversity (herb, shrub, and tree layers combined) relative to the control. However, differences in total structural diversity between treatments and control were, for the most part, not significant ($P > 0.10$). Therefore, these vegetation management treatments affected only the volume of the targeted deciduous tree layer and did not adversely affect species richness, diversity, turnover, or structural diversity of the plant community. These results may be applicable to other temperate forest ecosystems where conifer release is practiced in young plantations (Lindgren and Sullivan 2001). Herb biomass and cover usually recover to untreated values within 2–3 yr of conifer release treatment (Sullivan 1994). Meanwhile, the reduced competition on target conifers allows enhanced growth with little adverse effect on plant diversity (Sullivan et al. 1996, 1998).

Nevertheless, some plant groups may take longer to recover from glyphosate application. For example, cryptogams (ferns) may take longer than 5 yr to recover in boreal forest situations (Newmaster and Bell 2002), probably reflecting longer generation times and poor dispersal. Reviewing the effects of glyphosate use in forestry, Sullivan and Sullivan (2003) noted that single applications of glyphosate control much of the vegetation that receives spray, but recovery is generally rapid and within the range of natural disturbances.

Overall, the experience of glyphosate use in northern temperate forests is that vegetation and fauna recover over 2–3 yr following a single conifer-release treatment. With generally rapid plant growth under tropical conditions, available data confirm this scenario for Colombian conditions.

In comparison, land clearance for agriculture, or coca/poppy production, is a much more environmentally damaging operation, impacting adversely on soils in particular. Land clearance for illicit crops is already a threat to the conservation of bird species diversity in Colombia (Álvarez 2002). Although there are legitimate scientific questions as to the effects of (a) the spatial scale of individual glyphosate applications and (b) the return frequency of eradication treatments, field operational factors set these parameters. Spray areas reflect the patch scale of coca and poppy growing, averaging 1–2 ha each in a total of ~150,000 ha. Reapplication frequencies are generally greater than 6 mon for coca and greater than 3 mon for poppy and, bearing in mind the molecule is biologically unavailable in the soil and soil-bound residues have a half-life of 14–32 d, the environmental impacts are no greater than single applications.

V. Risk Assessment

The risk assessment was conducted by comparing estimated exposures to effect values for glyphosate from specific toxicity studies, from the literature, and from regulatory guidelines such as those established by the USEPA (1993b). The estimated exposures used were those calculated for the use of glyphosate for eradication spraying in Colombia.

A. Human Health

From an assessment of the results of toxicity testing of the formulation of glyphosate and Cosmo-Flux as used in Colombia, it was concluded that the addition of Cosmo-Flux to the spray mixture did not affect toxicity of the glyphosate to mammals. For this reason, it was possible to compare the toxicity of glyphosate and its formulations to exposures estimated under conditions of use in Colombia.

Exposures for the assessment were taken from Tables 5–7. The greatest values were taken as reasonable worst case for a hazard assessment. These results are shown in Table 12 and illustrated in Fig. 9. In comparing the exposure and effect concentrations, a margin of exposure approach was used. Thus a number greater than 1 (in Table 12) means that the exposure was less than criterion value or the exposure (or dose) that caused the response in the toxicology study. From these data, it is clear that potential exposures to glyphosate and Cosmo-Flux do not present a risk to human bystanders. In all cases, the margin of exposure for the most sensitive endpoint in laboratory animal studies with glyphosate was greater than 100, a conservative value often used to account for uncertainty in risk assessments of this type. As well, estimated worst case exposures were below the reference dose (RfD) established for glyphosate by the USEPA. The toxicity values used in both these approaches were derived from chronic exposures where the animals were dosed over extended time periods. They are thus

Table 12. Summary of reasonable worst case estimated exposures of humans to glyphosate resulting from use in the eradication of coca and poppy in Colombia and margins of exposure.

	Exposure value (mg/kg)		Margin of exposure compared to the most sensitive NOEL (175 mg/kg bw)	
Source of exposure	Coca	Poppy	Coca	Poppy
Direct overspray	0.04	0.01	4,918	20,417
Reentry	0.26	0.06	676	2,804
Inhalation	0.01	0.01	28,226	28,226
Diet and water	0.75	0.18	234	972
Worst case total exposure from all sources	1.05	0.26	167	680

	Exposure value (mg/kg)		Margin of exposure for the U.S. EPA RfD (2 mg/kg bw/d)	
Source of exposure	Coca	Poppy	Coca	Poppy
Direct overspray	0.04	0.01	56	233
Reentry	0.26	0.06	8	32
Inhalation	0.01	0.01	323	323
Diet and water	0.75	0.18	2.7	11.1
Worst case total exposure from all sources	1.05	0.26	1.9	7.8

additionally protective of short and infrequent exposures that would occur during the use of glyphosate in the eradication spray program. Some exposure values were close to the inhalation toxicity value, but as already discussed, droplet size is large and inhalation will be less than in the laboratory animal studies as well as the droplet size used in agriculture, from which the potential inhalation exposure was derived.

B. Environment

The acute toxicity data for formulated glyphosate in aquatic animals from Solomon and Thompson (2003) were combined with some of the new data for amphibians described above and are displayed graphically as a point of reference for characterizing the toxicity of glyphosate plus Cosmo-Flux as

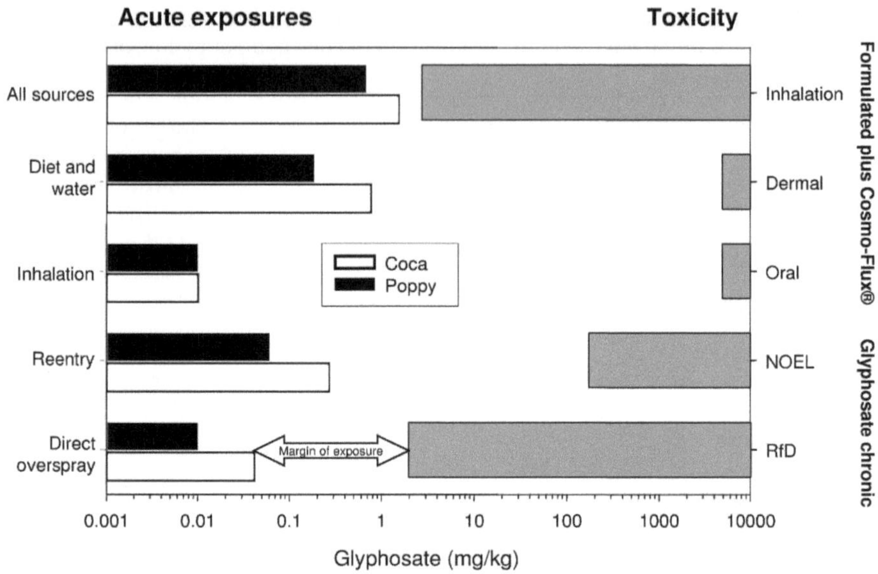

Fig. 9. Illustration of acute toxicity values in laboratory mammals for glyphosate plus Cosmo-Flux, the no-observed-effect-level (NOEL) from the most sensitive chronic study in laboratory animals, and the reference dose (RfD) (glyphosate) and the estimated worst case acute exposures that may be experienced under conditions of use in Colombia.

used in Colombia (Fig. 10). The graph is presented as a cumulative frequency distribution in a manner similar to that used in probabilistic risk assessments for pesticides (Solomon and Takacs 2002). The combination of formulated glyphosate and Cosmo-Flux, as used in Colombia, is more toxic to the aquatic organisms tested than formulations of glyphosate without the addition of surfactants and/or adjuvants, which is not altogether surprising. The toxicity of glyphosate itself to aquatic organisms is very small (Solomon and Thompson 2003) but, when mixed with some surfactants and adjuvants, this toxicity can be increased. The toxicity of Cosmo-Flux was not tested on its own; however, from experience with other adjuvants, it clearly contributes to the increased toxicity of the mixture. It is interesting to note that larval amphibians appear to be more susceptible to glyphosate formulations than other aquatic animals. The reason for this is likely the surfactants in the formulation of Roundup, as already discussed, as other formulations of glyphosate are less toxic to amphibians (Howe et al. 2004).

Assessment of the environmental risks of glyphosate and Cosmo-Flux to aquatic organisms was based on toxicity data from the literature and from studies conducted on the mixture of formulated glyphosate and Cosmo-Flux as used in Colombia. When the toxicity values for the mixture as used in Colombia are compared with the range of estimated exposures that would result from a direct overspray of surface waters (see Table 8), it is clear that aquatic animals and algae in some shallow water bodies may be at risk (see Fig. 10).

Although the overlap of the range of estimated exposure concentrations with the toxicity values for green alga and rainbow trout suggest that there may be increased risk in situations where an accidental overspray will occur, this would have to be in a location where a shallow water body is close enough to the coca field that it is accidentally oversprayed, that it is less than 30 cm deep, and that it is not flowing. Water flow would likely result in rapid hydraulic dilution to concentrations below the threshold of biological activity, so organisms in flowing water would not be at great risk. It was not possible to determine the actual frequency of these risks, as data on proximity of surface water to coca fields are not available. Based on the toxicity data with formulated Roundup in amphibians, this group of organisms may be at risk; however, specific testing in amphibians has not been conducted on the glyphosate plus Cosmo-Flux as used in Colombia.

Based on the toxicity data for honeybees, glyphosate and Cosmo-Flux is not acutely toxic via contact exposure to honeybees. It caused no mortality

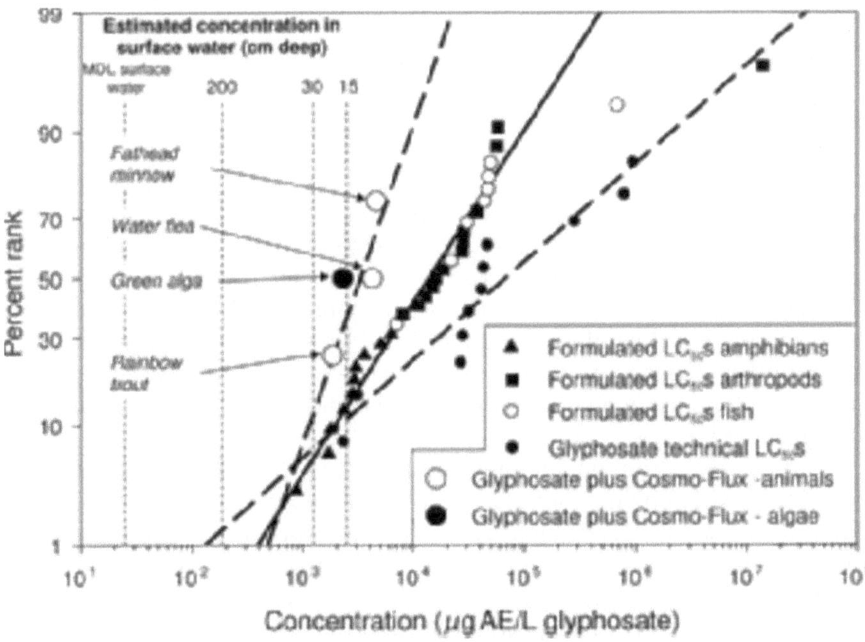

Fig. 10. Cumulative frequency distribution of toxicity values for glyphosate technical, formulated glyphosate (Roundup) in all aquatic organisms and in fish and the toxicity values in four aquatic species for glyphosate and Cosmo-Flux 411 mixture as used in Colombia.

or stress effects in bees in the normal 48-hr period after treatment at concentrations equal to or less than 63.9 mg AE/bee, showing that the formulated product is not directly hazardous to bees or, by extrapolation, to other beneficial insects.

Although no acute or chronic data are available on wild animals, extrapolation of the mammalian data discussed earlier and reports in the literature support the conclusion that glyphosate and Cosmo-Flux will not have adverse direct effects on wild mammals or birds. Indirect effects through habitat alteration are possible. However, it is unlikely that the coca and poppy fields are significant habitats for wildlife. Human activities related to cultivation, pest control, and harvesting will be more disruptive to wildlife, and death of the coca bushes or the poppy plants as a result of spraying with glyphosate will not add an additional stressor. In fact, if the sprayed area is not replanted and allowed to naturalize, this new successional habitat may be more attractive to birds and mammals than an old-growth forest. Given that coca and poppy fields are usually located in remote areas and are often surrounded by natural habitats, sources for recolonization or alternate habitats will be close by. Some habitat alteration will result from accidental oversprays that affect nontarget vegetation; however, as already discussed, these areas are small in relation to the sprayed fields (<0.48%), represent a very small proportion of the total habitat available (≪0.001%), and will undergo rapid recolonization and succession to habitats suitable for wildlife.

VI. Conclusions

Risks associated with the use of glyphosate and Cosmo-Flux in the coca and poppy eradication program in Colombia are related to the total impacts of coca and poppy production discussed in the section on Problem Formulation. There are a number of other activities associated with the production of cocaine and heroin that result in risks to human health and the environment. Data are not available to quantify all these risks, but some of them may be estimated on the basis of other knowledge and expert judgment, which was done using an adaptation of a risk prioritization scheme used in ecological risk assessment (Harwell et al. 1992).

For purposes of ranking human health hazards, the intensity score ranged from 0 to 5, with 5 being a severe effect such as a physical injury or toxicity. The recovery score also ranged from 0 to 5 and was based on the potential for complete recovery from the adverse effect. Frequency was based on an estimate of the proportion (%) of the total number of persons involved in coca and poppy cultivation, production, and the refinement of cocaine and heroin. The score for impact was the product of the individual scores and the percent impact is based on the sum of the impact scores.

A similar procedure to that described above was used for ranking ecological risks associated with the cycle of coca and poppy production.

The intensity score was ranked from 0 to 5, with 5 being most intense, such as the total destruction of the habitat by clear-cutting and burning a natural area. Intensity of effects in this case also included off-field effects such as on nontarget animals and plants. Recovery time in this scheme is the estimated time for the impacted area to recover to a state similar to the initial condition. In the case of clear-cutting and burning, it is recognized that succession will begin immediately; however, full recovery to a mature and diverse tropical forest may take considerably more than the 60 yr estimated here. Similarly, in the absence of cultivation, it was estimated that invasive and competitive species will displace coca and poppy in several years, and an estimate of 4 yr was used in this case. Given the need to apply fertilizer and pesticides frequently because of utilization of nutrients and resurgence of pests, the recovery time for these ecological impacts was judged to be small. The scores were multiplied to give the impact score, and the percent impact was based on the sum of the impact scores.

A. Human Health Relevance

Based on all the evidence and information presented here, we concluded that the risks to humans and human health from the use of glyphosate and Cosmo-Flux in the control of coca and poppy were minimal (Table 13). Acute toxicity of the formulated product and Cosmo-Flux to laboratory animals was very low, the likely exposures were low, and the frequency of exposures was low. When these risks are compared with other risks associated with clearing of land, the uncontrolled and unmonitored use of other pesticides (many of them more toxic to humans than glyphosate, CICAD/

Table 13. Potential human health impacts of the cycle of coca or poppy production in the Colombian environment.

Impacts	Intensity score	Recovery score	Frequency (%)	Impact score	Impact (%)
Clear cutting and burning	5	3	3	45	16.7
Planting the coca or poppy	0	1	100	0	0.0
Fertilizer inputs	0	0.5	10	0	0.0
Pesticide inputs	5	3	10	150	55.6
Eradication spray	0	0	10	0	0.0
Processing and refining	5	3	5	75	27.8

OAS 2005) to protect the coca and poppy, and exposures to substances used in the refining of the raw product into cocaine and heroin, they are essentially negligible.

B. Ecological Relevance

Based on evidence and data discussed above and results of a number of specific studies conducted specifically for this assessment, we concluded that the risks to the environment from the use of glyphosate and Cosmo-Flux in the control of coca and poppy were small in most circumstances (Table 14). Risks of direct effects in terrestrial wildlife such as mammals and birds were judged to be negligible, as were those to beneficial insects such as bees. Moderate risks to some aquatic wildlife may exist in some locations where shallow and static water bodies are located in close proximity to coca fields and are accidentally oversprayed. However, when taken in the context of the environmental risks from other activities associated with the production of coca and poppy, in particular, the uncontrolled and unplanned clearing of pristine lands in ecologically important areas for the purposes of planting the crop, the added risks associated with the spray program are small.

Table 14. Potential environmental impacts of the coca or poppy production cycle in the Colombian environment.

Impacts	Intensity score	Recovery time (years)	Impact score	Impact%
Clear cutting and burning	5	60	300	96.9
Planting the coca or poppy	1	4	4	1.3
Fertilizer inputs	1	0.5	0.5	0.2
Pesticide inputs	5	0.5	2.5	0.8
Eradication spray	1	0.5	0.5	0.2
Processing and refining	2	1	2	0.6

C. Strengths and Uncertainties in the Assessment

This assessment has both strengths and uncertainties, as discussed in the following sections; these lie in the exposure and effects characterizations and, because they are used in the risk characterization, are also reflected in the risk assessment. Uncertainties are inherent in all risk assessments and, in some cases, can be easily addressed though additional data collection or specific studies. Recommendations for additional studies and data collection are addressed in the final section.

Exposures

Human Exposures. Human exposures to glyphosate were estimated from extensive and well-documented studies in other jurisdictions and are judged to be accurate with respect to bystanders who are directly oversprayed. Exposures were judged to be small and, in all cases, considerably below thresholds of concern.

Application rates of glyphosate used for coca control are greater than those used in conventional agriculture, suggesting that experience and exposures measured under these conditions may not be applicable to bystander exposures in eradication spraying in Colombia. While this may be true, the margins between exposures doses at which chronic effects may occur are great enough to provide a wide margin of safety to bystanders. Less information is available regarding the likelihood of exposure upon reentry to coca fields immediately after spraying; this relates to the anecdotal evidence that picking of leaves or pruning of plants immediately after they are sprayed with glyphosate will "save" the plants. Exposures under these conditions are unmeasured, but are estimated to be below the USEPA reference dose.

Environmental Exposures. Applications of glyphosate are well characterized using state-of-the-art equipment. Locations of application and areas sprayed are well documented and measured with resolutions only equaled in some applications in forestry in other jurisdictions. Mixing and application rates are well characterized, and the probability of application of amounts of glyphosate and Cosmo-Flux greater than those specified is small. Concentrations in soil and water that may result from an accidental overspray also have high certainty. The environmental behavior of glyphosate is well characterized and, under the conditions of use, it will not persist, accumulate, or biomagnify. Analyses of surface waters and sediments in one watershed where eradication spraying was carried out did not reveal the presence of significant concentrations of glyphosate, confirming the conclusion based on its properties that it is not mobile in the environment. Residues of glyphosate were infrequently detected in areas where eradication spraying was not conducted but where glyphosate use was known to occur in agriculture. Given that considerably more glyphosate is used in agriculture and other noneradication uses (~85%), this further confirms that glyphosate is not sufficiently mobile to result in significant contamination of surface waters in Colombia, regardless of the use pattern.

Uncertainties in the exposure characterization lie in lack of precise measurements of the proximity of sprayed fields to surface waters and the proportion of treated areas that are in close proximity. Sampling of the surface waters only took place for 24 wk, and only five locations were sampled. Although two of these were scheduled to be sprayed, only one was treated during the sampling period. For logistical reasons, it was also not possible to sample close to application sites. Had sampling been conducted at more sites closer to the sprayed fields and over a longer time period, residues may have been detected more frequently.

Effects

Effects in Humans. The database of glyphosate effects is large and its risks to humans and the environment have been extensively reviewed and assessed in a number of national and international jurisdictions as well as in the open scientific literature. In all cases, glyphosate poses little risk. However, some of the studies on which these assessments are based were conducted before the refinement of testing guidelines and the availability of new and more sensitive methods of analysis and effect characterization, such as those based on alteration in the concentrations of neurotransmitters and their metabolites in the central nervous system. In the process of reassessment and reregistration, older studies will be replaced with newer tests using current guidelines. Given the large and expanding use of glyphosate in agriculture, priorities for updating the database will likely be high. Changes in the regulatory status of glyphosate should be monitored and any newly identified risks included in an updated risk assessment.

There is considerable literature on the epidemiology of pesticides and possible effects on human health. As a result of recent work, it is clear that many epidemiology studies are confounded by the use of poor and inaccurate surrogates for exposures to pesticides. We also conducted a preliminary epidemiological study to assess possible linkages between the use of glyphosate and adverse human health outcomes; this study recognizes that, for clear logistical reasons, no measures of exposure were available for the various groups enrolled in the study, other than the use of glyphosate for eradication spraying in the region. The results do not suggest that there is an association between the use of glyphosate in the eradication program and time to pregnancy (TTP) as a reproductive outcome. A somewhat greater risk for longer TTP was observed in one region (Valle del Cauca) where eradication spraying is not conducted, but it was not possible to identify any specific factors that may have been responsible for this observation.

Environmental Effects. The environmental toxicology database for glyphosate is relatively large, and its effects in nontarget organisms are well known or can be extrapolated. Glyphosate itself is essentially nontoxic to nontarget organisms. However, there are a number of formulations of glyphosate in the marketplace that contain many different surfactants and/or adjuvants. It is also known that it is the surfactants that determine the toxicity of the formulation and are more toxic than technical glyphosate itself. Because of this, several toxicity tests were conducted with the formulated product of glyphosate plus Cosmo-Flux used in the Colombian program; this reduced uncertainty with respect to toxicity to beneficial insects, such as the honeybee, and to aquatic organisms. Recent studies have reported that frogs are among the more-sensitive aquatic organisms to formulations of glyphosate such as Roundup and Vision. We did not conduct toxicity studies in amphibians with the mixture of glyphosate plus Cosmo-Flux, and this is a source of some uncertainty for ecological risks for frogs.

Confounding Risks

Through the Tier 1 and Tier 2 hazard assessments of the other substances used in the production and refining of cocaine and heroin (CICAD/OAS 2004c, 2005), we recognize that some of these substances present a significantly greater hazard to both humans and the environment than does the mixture of glyphosate and Cosmo-Flux used in the program. Exacerbating these hazards is the lack of information about the conditions of their use. Because of the lack of specific data on use and exposure, it was not possible to conduct detailed risk assessments for these substances. From anecdotal evidence and observations in other locations, it is clear that, in most cases, these substances are used without adequate safety training, without adequate protective equipment, without suitable disposal methods, and without supervision, which represents a significant and serious potential risk to humans and the environment.

D. Recommendations

We have identified a number of uncertainties in our review of the data and, from these, make the following recommendations. These recommendations are grouped into two classes, recommendations to retain current practices that were judged to be essential or useful (Table 15) and recommendations related to new activities or data collection which will address key uncertainties identified in our study (Table 16). As already noted, risk assessments require review and reevaluation from time to time. Thus, our recommendations include the updating of this risk assessment as additional data become available.

Summary

The production of coca and poppy as well as the processing and production of cocaine and heroin involve significant environmental impacts. Both coca and poppy are grown intensively in a process that involves the clearing of land in remote areas, the planting of the crop, and protection against pests such as weeds, insects, and pathogens. The aerial spray program to control coca and poppy production in Colombia with the herbicide glyphosate is

Table 15. Recommendations for the continuance of current practices in the coca and poppy eradication program in Colombia.

Practice	Benefit of continuance	Ranking of importance (5 = most important)
Mixer-loader, worker, and environmental protection in the storage, mixing, and loading operations.	Protection of the humans and the environment from excessive exposures.	5
Use of state-of-art application technology.	Accurate records of location and areas sprayed.	5
Replace the respirator worn by the mixer-loader with a full face shield to reduce the potential for splashed material to run down the face into the eyes.	This recommendation is procedures modification of current that will reduce the risk of splashes of concentrated formulation into the eyes.	5
Use of glyphosate in the eradication program.	The risk of this product to humans and the environment is judged to be smaller than any currently available alternatives. However, if new candidate products become available, their use should be considered only after an appropriate risk assessment has been conducted.	4

conducted with modern state-of-the-art aircraft and spray equipment. As a result of the use of best available spray and navigation technology, the likelihood of accidental off-target spraying is small and is estimated to be less than 1% of the total area sprayed.

Estimated exposures in humans resulting from direct overspray, contact with treated foliage after reentry to fields, inhalation, diet, and drinking water were small and infrequent. Analyses of surface waters in five watersheds showed that, on most occasions, glyphosate was not present at measurable concentrations; only two samples had residues just above the method detection limit of 25 µg/L. Concentrations of glyphosate in air were predicted to be very small because of negligible volatility. Glyphosate in soils that are directly sprayed will be tightly bound and biologically unavailable and have no residual activity. Concentrations of glyphosate plus Cosmo-Flux will be relatively large in shallow surface waters that are directly

Table 16. Recommendations for the collection of new data and information in the coca and poppy eradication program in Colombia.

Recommendation	Benefit of new data	Ranking of importance (5 = most important)
Conduct a study to identify risk factors associated with time to pregnancy (TTP).	This is a recommendation resulting from the observation of increased risk of longer TTP in one region of Colombia (Valle del Cauca) where eradication spraying was not carried out. The study should be considered for prioritization in the general human health research programs conducted in Colombia.	3
Including proximity to surface waters in Geographic Information System (GIS) analysis of locations and areas of coca and poppy fields.	Better indication of likely frequency of contamination of these habitats; this would help to better quantify the risks to aquatic organisms in shallow-water nonflowing habitats.	2
Identify mixtures of glyphosate and adjuvants that are less toxic to aquatic organisms than the currently used mixture. The priority of this recommendation would depend on the results of the GIS analysis.	Reduction in possible environmental impacts to nontarget organisms in shallow surface water environments.	2
Testing of the glyphosate-Cosmo-Flux formulation for toxicity to amphibians.	Decrease in uncertainty regarding the toxicity to amphibians.	2
Use of GIS to quantify areas of coca and poppy production in biodiversity hotspots.	Better quantification of proportion of regions identified as important sources of biodiversity that are being adversely impacted because of clear-cutting and planting of coca and poppy.	2

Table 16. *Continued*

Recommendation	Benefit of new data	Ranking of importance (5 = most important)
Use of GIS to quantify size of fields planted to coca and poppy and track these over time to judge extent of environmental impact as well as recovery.	Allow more-accurate quantification of potentially impacted areas as well as recovery when these fields are abandoned.	2
Review the regulatory status of glyphosate on a regular basis.	Ensure that new testing and toxicity data on glyphosate are included in the risk assessment of its use in eradication spraying in Colombia.	2
Measurement of exposures to glyphosate in bystanders to sprays and reentry into sprayed fields. This recommendation would follow selection of new formulations and mixtures of adjuvants that have less environmental toxicity.	Better characterization of exposures under conditions of use in Colombia.	1

oversprayed (maximum instantaneous concentration of 1,229 μg AE/L in water 30 cm deep); however, no information was available on the number of fields in close proximity to surface waters, and thus it was not possible to estimate the likelihood of such contamination.

The formulation used in Colombia, a mixture of glyphosate and Cosmo-Flux, has low toxicity to mammals by all routes of exposure, although some temporary eye irritation may occur. Published epidemiological studies have not suggested a strong or consistent linkage between glyphosate use and specific human health outcomes. An epidemiology study conducted in Colombia did not show any association between time to pregnancy in humans and the use of glyphosate in eradication spraying.

The mixture of glyphosate and Cosmo-Flux was not toxic to honeybees. The mixture was, however, more toxic to the alga *Selenastrum*, the cladoceran *Daphnia magna*, fathead minnow, and rainbow trout than formulated glyphosate (Roundup) alone. Studies on the use of glyphosate in agriculture and forestry have shown that direct effects on nontarget organisms other than plants are unlikely. Indirect effects on terrestrial arthropods and other wildlife may be the result of habitat alteration and environmental change brought about by the removal of plants by glyphosate. Because of the lack of residual activity, recovery of glyphosate-treated areas in Colombia is expected to be rapid because of good plant growth conditions. However, return to the conditions of tropical old-growth forest that existed before clear-cutting and burning may take hundreds of years, not from the use of glyphosate but because of the clear-cutting and burning, which are the primary cause of effects in the environment.

The risk assessment concluded that glyphosate and Cosmo-Flux did not present a significant risk to human health. In the entire cycle of coca and poppy production and eradication, human health risks associated with physical injury during clear-cutting and burning and the use of pesticides for protection of the illicit crops were judged to be considerably more important than those from exposure to glyphosate. For the environment, direct risks from the use of glyphosate and Cosmo-Flux to terrestrial mammals and birds were judged to be negligible. Moderate risks could occur in aquatic organisms in shallow surface waters that are oversprayed during the eradication program. However, the frequency of occurrence and extent to which this happens are unknown as data on the proximity of surface waters to coca fields were not available. Considering the effects of the entire cycle of coca and poppy production and eradication, clear-cutting and burning and displacement of the natural flora and fauna were identified as the greatest environmental risks and are considerably more important than those from the use of glyphosate for the control of coca and poppy.

Acknowledgments

This review is based on a report (Solomon et al. 2005a) prepared for the Inter-American Drug Abuse Control Commission (CICAD) section of the Organization of American States (OAS) in response to requests from the Governments of Colombia, the United Kingdom, and the United States. The request was to conduct a science-based risk assessment of the human health and environmental effects of the herbicide, glyphosate, used for the control of the illicit crops coca and poppy in Colombia.

With an international panel of experts and activities in several countries, a study of this nature requires good coordination and organization. We are deeply indebted to Mr. Jorge Rios and Ms. Adriana Henao of the CICAD office for their excellent work in organizing meetings, teleconferences, and field trips. They served the authors well and frequently worked well beyond the call of duty. We are also very grateful for the contributions of the Colombian Field Team. Unfortunately, we cannot name these individuals; however, we extend our most grateful thanks to all of you for all the hard work and the personal risks that you took on behalf of data collection for this project.

Field visits to Colombia by the authors were facilitated and coordinated by the staff of the Ministry of Foreign Affairs, and we were afforded protection by the National Police (Antinarcoticos). We offer our grateful thanks to Brigadier General Luis Gómez, his staff, the pilots, technicians, and the "Junglas" commandos for aiding us in our observations and sampling and for tolerating our scientific curiosity in the face of other priorities. At all times, we were given free and unfettered access to information, were allowed to take photographs freely, and were always treated with respect and in a most professional manner.

The panel members are indebted to Drs. Lesbia Smith, Angus Crossan, Richard Brain, and also to the many students in the Toxicology Program at the University of Guelph for their work on the separate reports on Tier 1 and Tier 2 hazard assessment of other substances used in the production and refining of cocaine and heroin. These data are presented in separate reports (CICAD/OAS 2004c, 2005).

References

Acquavella JF, Weber JA, Cullen MR, Cruz OA, Martens MA, Holden LR, Riordan S, Thompson M, Farmer D (1999) Human ocular effects from self-reported exposures to Roundup® herbicides. Hum Exp Toxicol 18:479–486.

Acquavella JF, Alexander BH, Mandel JS, Gustin C, Baker B, Chapman P, Bleeke M (2004) Glyphosate biomonitoring for farmers and their families: results from the farm family exposure study. Environ Health Perspect 112:321–326.

Adler GH, Arboledo JJ, Travi BL (1997) Diversity and abundance of small mammals in degraded tropical dry forest of northern Colombia. Mammalia 61: 361–370.

Álvarez MD (2002) Illicit crops and bird conservation priorities in Colombia. Conserv Biol 16:1086–1096.

Amerio P, Motta A, Toto P, Pour SM, Pajand R, Feliciani C, Tulli A (2004) Skin toxicity from glyphosate-surfactant formulation. J Toxicol Clin Toxicol 42: 317–319.

Araujo AS, Monteiro RT, Abarkeli RB (2003) Effect of glyphosate on the microbial activity of two Brazilian soils. Chemosphere 52:799–804.

Arbuckle TE, Lin Z, Mery LS (2001) An exploratory analysis of the effect of pesticide exposure on the risk of spontaneous abortion in an Ontario farm population. Environ Health Perspect 109:851–857.

Arbuckle TE, Cole DC, Ritter L, Ripley BD (2004) Farm children's exposure to herbicides: comparison of biomonitoring and questionnaire data. Epidemiology 15:187–194.

Arbuckle T, Cole DC, Ritter L, Ripley BD (2005) Biomonitoring of herbicides in Ontario farm applicators. Scand J Work Environ Health 31(suppl 1):90–97.

Arregui MC, Lenardon A, Sanchez D, Maitre MI, Scotta R, Enrique S (2004) Monitoring glyphosate residues in transgenic glyphosate-resistant soybean. Pestic Manag Sci 60:163–166.

Bababurmi EA, Olorunsogo OO, Bassir O (1978) Toxicology of glyphosate in rats and mice. Toxicol Appl Pharmacol 45:319–320.

Baerson SR, Rodriguez DJ, Tran M, Feng Y, Biest NA, Dill GM (2002) Glyphosate-resistant goosegrass. Identification of a mutation in the target enzyme 5-enolpyruvylshikimate-3-phosphate synthase. Plant Physiol 129:1265–1275.

Barbosa ER, Leiros da Costa MD, Bacheschi LA, Scaff M, Leite CC (2001) Parkinsonism after glycine-derivate exposure. Movement Disord 16:565–567.

Battaglin WA, Kolpin DW, Scribner EA, Kuivila KM, Sandstrom MW (2005) Glyphosate, other herbicides, and transformation products in midwestern streams, 2002. J Am Water Res Assoc 41:323–332.

BC Ministry of Forests (2000) Boreal plant community diversity 10 years after glyphosate treatment: a report summary. Silviculture Practices Section, Victoria, BC, Canada.

BCPC (2003) The e-Pesticide Manual. British Crop Protection Council, Farnham, Surrey, UK.

Bell FW, Lautenschlager RA, Wagner RG, Pitt DG, Hawkins JW, Ride KR (1997) Motor-manual, mechanical, and herbicide release affect early successional vegetation in northwestern Ontario. For Chronol 73:61–68.

Bell JR, Johnson PJ, Hambler C, Haughton AJ, Smith H, Feber RE, Tattersall FH, Hart BH, Manley W, Macdonald DW (2002) Manipulating the abundance of *Lepthyphantes tenuis* (Araneae: Linyphiidae) by field margin management. Agric Ecosyst Environ 93:295–304.

Benedetti AL, Vituri CD, Trentin AG, Domingues MAC, Alvarez-Silva M (2004) The effects of sub-chronic exposure of Wistar rats to the herbicide Glyphosate-Biocarb (R). Toxicol Lett 153:227–232.

Bengtsson G, Hansson LA, Montenegro K (2004) Reduced grazing rates in *Daphnia pulex* caused by contaminants: implications for trophic cascades. Environ Toxicol Chem 23:2641–2648.

Blackburn LG, Boutin C (2003) Subtle effects of herbicide use in the context of genetically modified crops: a case study with glyphosate (Roundup (R)). Ecotoxicology 12:271–285.

Blakley BR (1997) Effect of Roundup and Tordon 202C herbicides on antibody production in mice. Vet Hum Toxicol 39:204–206.

Boahene K (1998) The challenge of deforestation in tropical Africa: reflections on its principal causes, consequences and solutions. Land Degrad Dev 9:247–258.

Boateng JO, Haeussler S, Bedford L (2000) Boreal plant community diversity 10 years after glyphosate treatment. West J Appl For 15:15–26.

Buckelew LD, Pedigo LP, Mero HM, Owen MDK, Tylka GL (2000) Effects of weed management systems on canopy insects in herbicide-resistant soybeans. J Econ Entomol 93:1437–1443.

Burgat V, Keck G, Guerre P, Pineau X (1998) Glyphosate toxicosis in domestic animals: a survey from the data of the Centre National d'Informations Toxicologiques Veterinaires (CNITV). Vet Hum Toxicol 40:363–367.

Campbell A, Chapman M (2000) Handbook of Poisoning in Dogs and Cats. Blackwell, Oxford.

Carlson AM, Gorchov DL (2004) Effects of herbicide on the invasive biennial *Alliaria petiolata* (garlic mustard) and initial responses of native plants in a southwestern Ohio forest. Restor Ecol 12:559–567.

Cauble K, Wagner RS (2005) Sublethal effects of the herbicide glyphosate on amphibian metamorphosis and development. Bull Environ Contam Toxicol 75:429–435.

Centre for Biodiversity (2004) Andean Biodiversity Region. Centre for Biodiversity. Accessed 2004 December 30. http://www.biodiversityhotspots.org/xp/Hotspots/andes/.
Chang CY, Peng YC, Hung DZ, Hu WH, Yang DY, Lin TJ (1999) Clinical impact of upper gastrointestinal tract injuries in glyphosate-surfactant oral intoxication. Hum Exp Toxicol 18:475–478.
Chen CY, Hathaway KM, Folt CL (2004) Multiple stress effects of Vision herbicide, pH, and food on zooplankton and larval amphibian species from forest wetlands. Environ Toxicol Chem 23:823–831.
Chinea JD (2002) Tropical forest succession on abandoned farms in the Humacao Municipality of eastern Puerto Rico. For Ecol Manag 167:195–207.
CICAD/OAS (2004a) Standard Operating Procedure. Stream Water and Sediment Sampling for Glyphosate/AMPA. CICAD, Organization of American States, Washington, DC.
CICAD/OAS (2004b) Standard Operating Procedure. Stream Water and Sediment Sampling for Other Pesticides. CICAD, Organization of American States, Washington, DC.
CICAD/OAS (2004c) The Toxicology of Chemicals Used in the Production and Refining of Cocaine and Heroin: A Tier-One Assessment. OAS/CICAD 2004-01. CICAD, Organization of American States, Washington, DC.
CICAD/OAS (2005) The Toxicology of Selected Chemicals Used in the Production and Refining of Cocaine and Heroin: A Tier-Two Assessment. OAS/CICAD 2005-01. CICAD, Organization of American States, Washington, DC. http://www.cicad.oas.org/Desarrollo_Alternativo/ENG/Projects%20By%20Country/Colombia/OAS_CICAD_Tier_2_Hazard_Assessment_July_2005%5B1%5D.pdf.
Clement RM, Horn SP (2001) Pre-Columbian land-use history in Costa Rica: a 3000-year record of forest clearance, agriculture and fires from Laguna Zoncho. Holocene 11:419–426.
Compton JE, Boone RD (2000) Long-term impacts of agriculture on soil carbon and nitrogen in New England forests. Ecology 81:2314–2330.
Conners DE, Black MC (2004) Evaluation of lethality and genotoxicity in the freshwater mussel *Utterbackia imbecillis* (Bivalvia: Unionidae) exposed singly and in combination to chemicals used in lawn care. Arch Environ Contam Toxicol 46:362–371.
Cosmoagro (2004) Cosmo-Flux® 411F Label. Cosmoagro S.A. Accessed October 20, 2004. www.cosmoagro.com.
Curtis JH, Brenner M, Hodell DA, Balser RA, Islebe GA, Hooghiemstra H (1998) A multi-proxy study of Holocene environmental change in the Maya lowlands of Peten, Guatemala. J Paleolimnol 19:139–159.
Curtis KM, Savitz DA, Weinberg CR, Arbuckle TE (1999) The effect of pesticide exposure on time to pregnancy. Epidemiology 10:112–117.
CWQG (1999) Canadian Water Quality Guidelines (and updates). Task Force on Water Quality Guidelines of the Canadian Council of Resource and Environment Ministers, Ottawa, ON. WWW.EC.GC.CA/CEQG-RCQE/WATER/HTM#TABL.
da Silva RS, Cognato GD, Vuaden FC, Rezende MFS, Thiesen FV, Fauth MD, Bogo MR, Bonan CD, Dias RD (2003) Different sensitivity of Ca^{2+}-ATPase and cholinesterase to pure and commercial pesticides in nervous ganglia of *Phyllocaulis soleiformis* (Mollusca). Comp Biochem Physiol C 135:215–220.

Dallegrave E, Mantese FD, Coelho RS, Pereira JD, Dalsenter PR, Langeloh A (2003) The teratogenic potential of the herbicide glyphosate-Roundup (R) in Wistar rats. Toxicol Lett 142:45–52.

Daruich J, Zirulnik F, Gimenez MS (2001) Effect of the herbicide glyphosate on enzyme activity in pregnant rats and their fetuses. Environ Res A 85:226–231.

De Roos AJ, Blair A, Rusiecki JA, Hoppin JA, Svec M, Dosemeci M, Sandler DP, Alavanja MC (2005) Cancer incidence among glyphosate-exposed pesticide applicators in the agricultural health study. Environ Health Perspect 113: 49–54.

Devine MD, Duke SO, Fedtke C (1993) Physiology of Herbicide Action. PTR Prentice Hall, Englewood Cliffs, NJ.

Diaz G, Carrillo C, Honrubia M (2003) Differential responses of ectomycorrhizal fungi to pesticides in pure culture. Cryptogam Mycol 24:199–211.

Dirección Nacional de Estupefacientes (2002) La lucha de Colombia contra las drogas ilicitas: acciones y resultados 2002. Colombia's war against drugs: actions and results 2002. Direccion Nacional de Estupefacientes, Bogota, D.C. Colombia.

Dyer SD, Belanger SE, Carr GJ (1997) An initial evaluation of the use of Euro/North America fish species for tropical effects assessments. Chemosphere 35:2767–2781.

Easton WE, Martin K (1998) The effect of vegetation management on breeding bird communities in British Columbia. Ecol Appl 8:1092–1103.

Easton WE, Martin K (2002) Effects of thinning and herbicide treatments on nest-site selection by songbirds in young, managed forests. Auk 119:685–694.

Edginton AN, Sheridan PM, Stephenson GR, Thompson DG, Boermans HJ (2004) Comparative effects of pH and Vision herbicide on two life stages of four anuran amphibian species. Environ Toxicol Chem 23:815–822.

El-Demerdash FM, Yousef MI, Elagamy EI (2001) Influence of paraquat, glyphosate, and cadmium on the activity of some serum enzymes and protein electrophoretic behaviour (in vitro). J Environ Sci Health B 36:29–42.

Ellis JM, Griffin JL, Linscombe SD, Webster ER (2003) Rice (*Oryza sativa*) and corn (*Zea mays*) response to simulated drift of glyphosate and glufosinate. Weed Technol 17:452–460.

Eschenburg S, Kabsch W, Healy ML, Schonbrunn E (2003) A new view of the mechanisms of UDP-*N*-acetylglucosamine enolpyruvyl transferase (MurA) and 5-enolpyruvylshikimate-3-phosphate synthase (AroA) derived from X-ray structures of their tetrahedral reaction intermediate states. J Biol Chem 278:49215–49222.

Eschholz WE, Servello FA, Griffith B, Raymond KS, Krohn WB (1996) Winter use of glyphosate-treated clearcuts by moose in Maine. J Wildl Manag 60:764–769.

Everett KDE, Dickerson HW (2003) *Ichthyophthirius multifiliis* and *Tetrahymena thermophila* tolerate glyphosate but not a commercial herbicidal formulation. Bull Environ Contam Toxicol 70:731–738.

Feng JC, Thompson DG (1990) Fate of glyphosate in in a Canadian forest watershed. 2 Persistence in foliage and soils. J Agric Food Chem 38:1118–1125.

Ferreira JFS, Reddy KN (2000) Absorption and translocation of glyphosate in *Erythroxylum coca* and *E. novogranatense*. Weed Sci 48:193–199.

Ferreira JFS, Smeda RJ, Duke SO (1997) Control of coca plants (*Erythroxylum coca* and *E. novogranatense*) with glyphosate. Weed Sci 45:551–556.

Fomsgaard IS, Spliid NH, Felding G (2003) Leaching of pesticides through normal-tillage and low-tillage soil—a lysimeter study. II. Glyphosate. J Environ Sci Health B 38:19–35.

Frank R, Campbell RA, Sirons GJ (1985) Forestry workers involved in aerial application of 2,4-dichlorophenoxyacetic acid (2,4-D): exposure and urinary excretion. Arch Environ Contam Toxicol 14:427–435.

Franz JE, Mao MK, Sikorski JA (1997) Glyphosate: A Unique Global Herbicide. ACS Monograph No. 189. American Chemical Society, Washington, DC.

Garry VF, Harkins ME, Erickson LL, Long-Simpson LK, Holland SE, Burroughs BL (2002) Birth defects, season of conception, and sex of children born to pesticide applicators living in the Red River Valley of Minnesota, USA. Environ Health Perspect 110(suppl 3):441–449.

Gauriguata MR, Ostertag R (2001) Neotropical secondary forest succession: changes in structural and functional characteristics. For Ecol Manag 148: 185–206.

Giesy JP, Dobson S, Solomon KR (2000) Ecotoxicological risk assessment for Roundup® herbicide. Rev Environ Contam Toxicol 167:35–120.

Goldstein DA, Acquavella JF, Mannion RM, Farmer DR (2002) An analysis of glyphosate data from the California Environmental Protection Agency Pesticide Illness Surveillance Program. J Clin Toxicol 40:885–892.

Goman M, Byrne R (1998) A 5000-year record of agriculture and tropical forest clearance in the Tuxtlas, Veracruz, Mexico. Holocene 8:83–89.

Gomez MA, Sagardoy MA (1985) Effect of glyphosate herbicide on the microflora and mesofauna of a sandy soil in a semiarid region. Rev Latinoam Microbiol 27:351–357.

Granby K, Vahl M (2001) Investigation of the herbicide glyphosate and the plant growth regulators chlormequat and mepiquat in cereals produced in Denmark. Food Addit Contam 18:898–905.

Haney RL, Senseman SA, Hons FM (2002) Effect of Roundup Ultra on microbial activity and biomass from selected soils. J Environ Qual 31:730–735.

Hardell L, Eriksson M (1999) A case-control study of non-Hodgkin lymphoma and exposure to pesticides. Cancer (Phila) 85:1353–1360.

Hardell L, Eriksson M, Nordstrom M (2002) Exposure to pesticides as risk factor for non-Hodgkin's lymphoma and hairy cell leukemia: pooled analysis of two Swedish case-control studies. Leuk Lymphoma 43:1043–1049.

Harris SA, Sass-Kortsak AM, Corey PN, Purdham J (2002) Development of models to predict dose of pesticides in professional turf applicators. J Expos Anal Environ Epidemiol 12:130–144.

Hartman KM, McCarthy BC (2004) Restoration of a forest understory after the removal of an invasive shrub, Amur honeysuckle (*Lonicera maackii*). Restor Ecol 12:154–165.

Harwell MA, Cooper W, Flaak R (1992) Prioritizing ecological and human welfare risks from environmental stresses. Environ Manag 16:451–464.

Haughton AJ, Bell JR, Wilcox A, Boatman ND (2001) The effect of the herbicide glyphosate on non-target spiders: Part I. Direct effects on *Lepthyphantes tenuis* under laboratory conditions. Pestic Manag Sci 57:1033–1036.

Helling CS (2003) Eradication of coca in Colombia—2002. Results of the Colombia coca verification mission #9, December 2002. U.S. Department of Agriculture, Beltsville, MD.

Hernandez A, Garcia-Plazaola JI, Becerril JM (1999) Glyphosate effects on phenolic metabolism of nodulated soybean (*Glycine max* L. merr.). J Agric Food Chem 47:2920–2925.

Howe CM, Berrill M, Pauli BD, Helbing CC, Werry K, Veldhoen N (2004) Toxicity of glyphosate-based pesticides to four North American frog species. Environ Toxicol Chem 23:1928–1938.

Huang X, Pedersen T, Fischer M, White R, Young TM (2004) Herbicide runoff along highways. 1. Field observations. Environ Sci Technol 38:3263–3271.

Hurtig AK, Sebastian MS, Soto A, Shingre A, Zambrano D, Guerrero W (2003) Pesticide use among farmers in the Amazon Basin of Ecuador. Arch Environ Health 58:223–228.

ICA (2003) Comercialización de Plaguicidas 2000–2001. Instituto Colombiano Agropecuario, Bogota, Colombia.

Idrovo J, Sanin LH, Cole D, Chavarro J, Caceres H, Narváez J, Restrepo M (2005) Time to first pregnancy among women working in agricultural production. Int Arch Occup Environ Health 78:493–500.

Immunopharmos (2002a) Irritación dérmica primaria con Glifosato 5% + Cosmoflux 1% + agua 94%. Immunopharmos Ltda. Laboratorios, Cota, Cundinamarca, Colombia.

Immunopharmos (2002b) Irritación dérmica primaria con Glifosato 44% + Cosmoflux 1% + agua 55%. Immunopharmos Ltda. Laboratorios, Cota, Cundinamarca, Colombia.

Immunopharmos (2002c) Irritación ocular primaria con Glifosato 5% + Cosmoflux 1% + agua 94%. Immunopharmos Ltda. Laboratorios, Cota, Cundinamarca, Colombia.

Immunopharmos (2002d) Irritación ocular primaria con Glifosato 44% + Cosmoflux 1% + agua 55%. Immunopharmos Ltda. Laboratorios, Cota, Cundinamarca, Colombia.

Immunopharmos (2002e) Sensibilidad cutánea en cobayos con Glifosato 5% + Cosmoflux 1% + agua 94%. Immunopharmos Ltda. Laboratorios, Cota, Cundinamarca, Colombia.

Immunopharmos (2002f) Sensibilidad cutánea en cobayos con Glifosato 44% + Cosmoflux 1% + agua 55%. Immunopharmos Ltda. Laboratorios, Cota, Cundinamarca, Colombia.

Immunopharmos (2002g) Toxicidad inhalatoria aguda-CL50 con Glifosato 5% + Cosmoflux 1% + agua 94%. Immunopharmos Ltda. Laboratorios, Cota, Cundinamarca, Colombia.

Immunopharmos (2002h) Toxicidad inhalatoria aguda-CL50 con Glifosato 44% + Cosmoflux 1% + agua 55%. Immunopharmos Ltda. Laboratorios, Cota, Cundinamarca, Colombia.

Immunopharmos (2002i) Toxicidad oral aguda-DL50 con Glifosato 5% + Cosmoflux 1% + agua 94%. Immunopharmos Ltda. Laboratorios, Cota, Cundinamarca, Colombia.

Immunopharmos (2002j) Toxicidad oral aguda-DL50 con Glifosato 44% + Cosmoflux 1% + agua 55%. Immunopharmos Ltda. Laboratorios, Cota, Cundinamarca, Colombia.

Jackson RE, Pitre HN (2004a) Influence of Roundup Ready soybean and roundup ultra herbicide on *Geocoris punctipes* (Say) (Heteroptera: Lygaeidae) in the laboratory. J Entomol Sci 39:56–61.

Jackson RE, Pitre HN (2004b) Influence of Roundup Ready soybean production systems and glyphosate application on pest and beneficial insects in narrow-row soybean. J Entomol Sci 39:62–70.

Jiraungkoorskul W, Upatham ES, Kruatrachue M, Sahaphong S, Vichasri-Grams S, Pokethitiyook P (2003) Biochemical and histopathological effects of glyphosate herbicide on Nile tilapia (*Oreochromis niloticus*). Environ Toxicol 18: 260–267.

Kimball AJ, Hunter ML (1990) Intensive silviculture. In: Hunter ML (ed) Wildlife, Forests and Forestry. Prentice Hall, Englewood Cliffs, NJ, pp 200–234.

Kjaer J, Ullum M, Olsen P, Sjelborg P, Helweg A, Bügel Mogensen B, Plauborg F, Grant R, Fomsgaard IS, Brüsch W (2003) The Danish Pesticide Leaching Assessment Programme. Monitoring Results May 1999–June 2002. 3rd Report. Geological Survey of Denmark and Greenland, Copenhagen, Denmark. http://www.geus.dk/pesticidvarsling/report_monitoring-results-99–02-uk.htm.

Kjaer J, Olsen P, Ullum M, Grant R (2005) Leaching of glyphosate and aminomethylphosphonic acid from Danish agricultural field sites. J Environ Qual 34:609–620.

Kolpin DW, Thurman EM, Lee EA, Meyer MT, Furlong ET, Glassmeyer ST (2006) Urban contributions of glyphosate and its degradate AMPA to streams in the United States. Sci Total Environ 354:191–197.

Krieger RI (2001) Handbook of Pesticide Toxicology. Academic Press, New York.

Krueger-Mangold J, Sheley RL, Roos BD (2002) Maintaining plant community diversity in a waterfowl production area by controlling Canada thistle (*Cirsium arvense*) using glyphosate. Weed Technol 16:457–463.

Laatikainen T, Heinonen-Tanski H (2002) Mycorrhizal growth in pure cultures in the presence of pesticides. Microbiol Res 157:127–137.

Lautenschlager RA, Sullivan TP (2002) Effects of herbicide treatments on biotic components in regenerating northern forests. For Chronol 78:695–731.

Lee HL, Chen KW, Chi CH, Huang JJ, Tsai LM (2000) Clinical presentations and prognostic factors of a glyphosate-surfactant herbicide intoxication: a review of 131 cases. Acad Emerg Med 7:906–910.

Lin N, Garry VF (2000) In vitro studies of cellular and molecular developmental toxicity of adjuvants, herbicides, and fungicides commonly used in Red River Valley, Minnesota. J Toxicol Environ Health A 60:423–439.

Lindgren PMF, Sullivan TP (2001) Influence of alternative vegetation management treatments on conifer plantation attributes: abundance, species diversity, and structural diversity. Forest Ecol Manag 142:163–182.

Lindsay EA, French K (2004) The impact of the herbicide glyphosate on leaf litter invertebrates within Bitou bush, *Chrysanthemoides monilifera* ssp. *rotundata*, infestations. Pestic Manag Sci 60:1205–1212.

Liu C-M, McLean PA, Sookdeo CC, Cannon FC (1991) Degradation of the herbicide glyphosate by memebrs of family *Rhizobiaceae*. Appl Environ Microbiol 57:1799–1840.

LSD (2005) Trace organic analyses. Accessed 2005 January 10. http://www.uoguelph.ca/labserv/Services/services.html.

Maltby L, Blake NN, Brock TCM, Van den Brink PJ (2005) Insecticide species sensitivity distributions: the importance of test species selection and relevance to aquatic ecosystems. Environ Toxicol Chem 24:379–388.

Mann RM, Bidwell JR (1999) The toxicity of glyphosate and several glyphosate formulations to four species of Southwestern Australian frogs. Arch Environ Contam Toxicol 36:193–199.

Manthey FA, Nalewaja JD (1992) Relative wax solubility and phytotoxicity of oil to green foxtail [*Setaria viridis* (L.) Beauv.]. In: Foy CL (ed) Adjuvants for Agrichemicals. CRC Press, Boca Raton, FL, pp 463–471.

Mårtensson AM (1992) Effects of agrochemicals and heavy metals on fast-growing *Rhizobia* and their symbiosis in small-seeded legumes. Soil Biol Biochem 25:435–445.

Martins SV, Rodrigues RR (2002) Gap-phase regeneration in a semideciduous mesophytic forest, south-eastern Brazil. Plant Ecol 163:51–62.

Martins SV, Junior RC, Rodrigues RR, Gandolfi S (2004) Colonization of gaps produced by death of bamboo clumps in a semideciduous mesophytic forest in south-eastern Brazil. Plant Ecol 172:121–131.

Martinson KB, Sothern RB, Koukkari WL, Durgan BR, Gunsolus JL (2002) Circadian response of annual weeds to glyphosate and glufosinate. Chronobiol Int 19:405–422.

Matarczyk JA, Willis AJ, Vranjic JA, Ash JE (2002) Herbicides, weeds and endangered species: management of bitou bush (*Chrysanthemoides monilifera* ssp. *rotundata*) with glyphosate and impacts on the endangered shrub, *Pimelea spicata*. Biol Conserv 108:133–141.

Matlack GR (1997) Four centuries of forest clearance and regeneration in the hinterland of a large city. J Biogeogr 24:281–295.

May WE, Loeppky HA, Murrell DC, Myhre CD, Soroka JJ (2003) Preharvest glyphosate in alfalfa for seed production: effect on alfalfa seed yield and quality. Can J Plant Sci 83:189–197.

McAlister JJ, Smith BJ, Sanchez B (1998) Forest clearance: impact of land-use change on fertility status of soils from the Sao Francisco area of Niteroi, Brazil. Land Degrad Dev 9:425–440.

McDonald MA, Healey JR, Stevens PA (2002) The effects of secondary forest clearance and subsequent land-use on erosion losses and soil properties in the Blue Mountains of Jamaica. Agric Ecosyst Environ 92:1–19.

Miller RP, Martinson KB, Sothern RB, Durgan BR, Gunsolus JL (2003) Circadian response of annual weeds in a natural setting to high and low application rates of four herbicides with different modes of action. Chronobiol Int 20: 299–324.

Morjan WE, Pedigo LP (2002) Suitability of transgenic glyphosate-resistant soybeans to green cloverworm (Lepidoptera: Noctuidae). J Econ Entomol 95:1275–1280.

Motavalli PP, Kremer RJ, Fang M, Means NE (2004) Impact of genetically modified crops and their management on soil microbially mediated plant nutrient transformations. J Environ Qual 33:816–824.

Mueller TC, Massey JH, Hayes RM, Main CL, Stewart CN (2003) Shikimate accumulates in both glyphosate-sensitive and glyphosate-resistant horseweed (*Conyza canadensis* L. Cronq.). J Agric Food Chem 51:680–684.

Nakashima K, Yoshimura T, Mori H, Kawaguchi M, Adachi S, Nakao T, Yamazaki F (2002) Effects of pesticides on cytokines production by human peripheral blood mononuclear cells—fenitrothion and glyphosate (in Japanese with English abstract). Chudoku Kenkyu 15:159–165.

Neve P, Diggle AJ, Smith FP, Powles SB (2003a) Simulating evolution of glyphosate resistance in *Lolium rigidum* I: population biology of a rare resistance trait. Weed Res 43:404–417.

Neve P, Diggle AJ, Smith FP, Powles SB (2003b) Simulating evolution of glyphosate resistance in *Lolium rigidum* II: past, present and future glyphosate use in Australian cropping. Weed Res 43:418–427.

Newmaster SG, Bell FW (2002) The effects of silvicultural disturbances on cryptogam diversity in the boreal-mixedwood forest. Can J For Res 32:38–51.

Nijman V (1998) Habitat preference of Great Argus pheasant (*Argusianus argus*) in Kayan Mentarang National Park, east Kalimantan, Indonesia. J Ornitol 139:313–323.

Norsworthy JK (2004) Conventional soybean plant and progeny response to glyphosate. Weed Technol 18:527–531.

NRA (1996) NRA Special Review of Glyphosate. 96.1. National Registration Authority for Agricultural and Veterinary Chemicals, Canberra, Australia.

NRC (1986) Ecological Knowledge and Environmental Problem-Solving: Concepts and Case Studies. National Academy Press, Washington, DC.

NRC (1993) Issues in Risk Assessment. National Academy Press, Washington, DC.

OECD (1984a) OECD Guidelines for testing of chemicals. Method 201. Alga, growth inhibition test. Organisation for Economic Co-Operation and Development, Paris.

OECD (1984b) OECD Guidelines for testing of chemicals. Method 202. *Daphnia* species, acute immobilisation test and reproduction test. Organisation for Economic Co-Operation and Development, Paris.

OECD (1992) OECD Guidelines for testing of chemicals. Method 203. Fish acute toxicity test. Organisation for Economic Co-Operation and Development, Paris.

OECD (1998a) OECD Guidelines for testing of chemicals. Method #214. Honeybees, acute contact toxicity test. Organisation for Economic Co-Operation and Development, Paris.

OECD (1998b) The OECD Principles of Good Laboratory Practice. Environment Monograph No. 45. Organisation for Economic Co-Operation and Development, Paris.

Okwakol MJN (2000) Changes in termite (Isoptera) communities due to the clearance and cultivation of tropical forest in Uganda. Afr J Ecol 38:1–7.

Pariona W, Fredericksen TS, Licona JC (2003) Natural regeneration and liberation of timber species in logging gaps in two Bolivian tropical forests. For Ecol Manag 181:313–322.

Patzoldt WL, Tranel PJ, Hager AG (2002) Variable herbicide responses among Illinois waterhemp (*Amaranthus rudis* and *A. tuberculatus*) populations. Crop Prot 21:707–712.

Payne NJ (1993) Spray dispersal from aerial silvicultural glyphosate applications. Crop Prot 12:463–469.

Payne NJ, Feng JC, Reynolds PE (1990) Off-target deposits and buffer zones required around water for aerial glyphosate applications. Pestic Sci 30:183–198.

Pena-Claros M (2003) Changes in forest structure and species composition during secondary forest succession in the Bolivian Amazon. Biotropica 35:450–461.

Perez A, Kogan M (2003) Glyphosate-resistant *Lolium multiflorum* in Chilean orchard. Weed Res 43:12–19.
Pline WA, Edmisten KL, Wilcut JW, Wells R, Thomas J (2003) Glyphosate-induced reductions in pollen viability and seed set in glyphosate-resistant cotton and attempted remediation by gibberellic acid (GA(3)). Weed Sci 51:19–27.
Policia Nacional Dirección Antinarcóticos (2005) Maps showing areas in Colombia where coca and poppy are grown. Direccion Nacional de Estupefacientes, Bogota, DC, Colombia.
Popper KR (1979) Truth, Rationality and the Growth of Scientific Knowledge, Vol V. Klostermann, Frankfurt am Main.
PTG (2005a) Informe Boyacá. PGT, Bogota, Colombia.
PTG (2005b) Informe Nariño. PGT, Bogota, Colombia.
PTG (2005c) Informe Putumayo. PGT, Bogota, Colombia.
PTG (2005d) Informe Sierra Nevada. PGT, Bogota, Colombia.
PTG (2005e) Informe Valle de Cauca. PGT, Bogota, Colombia.
Pushnoy LA, Avnon LS, Carel RS (1998) Herbicide (Roundup) pneumonitis. Chest 114:1769–1771.
Quaghebeur D, De Smet B, De Wulf E, Steurbaut W (2004) Pesticides in rainwater in Flanders, Belgium: results from the monitoring program 1997–2001. J Environ Monit 6:182–190.
Racke KD, Skidmore MW, Hamilton DJ, Unsworth JB, Miyamoto J, Cohen SZ (1997) Pesticide fate in tropical soils. Pure Appl Chem 69:1349–1371.
Ramwell CT, Heather AI, Shepherd AJ (2002) Herbicide loss following application to a roadside. Pest Manag Sci 58:695–701.
Ramwell CT, Heather AI, Shepherd AJ (2004) Herbicide loss following application to a railway. Pest Manag Sci 60:556–564.
Raymond KS, Servello FA, Griffith B, Eschholz WE (1996) Winter foraging ecology of moose on glyphosate-treated clearcuts in Maine. J Wildl Manag 60: 753–763.
Reeves BG (1992) The rationale for adjuvant use with agrichemicals. In: Foy CL (ed) Adjuvants for Agrichemicals. CRC Press, Boca Raton, FL, pp 487–488.
Relyea RA (2004) Growth and survival of five amphibian species exposed to combinations of pesticides. Environ Toxicol Chem 23:1737–1742.
Relyea RA (2005a) The impact of insecticides and herbicides on the biodiversity and productivity of aquatic communities. Ecol Appl 15:618–627.
Relyea RA (2005b) The lethal impact of Roundup on aquatic and terrestrial amphibians. Ecol Appl 15:1118–1124.
Renz MJ, DiTomaso JM (2004) Mechanism for the enhanced effect of mowing followed by glyphosate application to resprouts of perennial pepperweed (*Lepidium latifolium*). Weed Sci 52:14–23.
Riley CM, Weisner CJ, Sexsmith WA (1991) Estimating off-target spray deposition on the ground following aerial application of glyphosate for conifer release in New Brunswick. J Environ Sci Health B 26:185–208.
Roberts CW, Roberts F, Lyons RE, Kirisits MJ, Mui EJ, Finnerty J, Johnson JJ, Ferguson DJ, Coggins JR, Krell T, Coombs GH, Milhous WK, Kyle DE, Tzipori S, Barnwell J, Dame JB, Carlton J, McLeod R (2002) The shikimate pathway and its branches in apicomplexan parasites. J Infect Dis 185(suppl 1):S25–S36.
Roberts F, Roberts CW, Johnson JJ, Kyle DE, Krell T, Coggins JR, Coombs GH, Milhous WK, Tzipori S, Ferguson DJ, Chakrabarti D, McLeod R (1998)

Evidence for the shikimate pathway in apicomplexan parasites. Nature (Lond) 393:801–805.
Rodwell DE (1980a) Teratology study in rabbits with technical glyphosate. 79/018. International Research and Development Corporation, Mattawan, MI. (Cited in Pesticide Residues in Food 1986. Evaluations 1986, Part II. Toxicology. Sponsored jointly by FAO and WHO, 1980.)
Rodwell DE (1980b) Teratology study in rats with technical glyphosate. 79/016. International Research and Development Corporation, Mattawan, MI. (Cited in Pesticide Residues in Food 1986. Evaluations 1986, Part II. Toxicology. Sponsored jointly by FAO and WHO, 1980.)
Rueppel ML, Brightwell BB, Schaefer J, Marvel JT (1977) Metabolism and degradation of glyphosate in soil and water. J Agric Food Chem 25:517–528.
Sanin L-H (2005) Analysis of Time to Pregnancy in Five Regions of Colombia, 2005. Autonomous University of Chihuahua and National Institute of Public Health, Mexico, Chihuahua, Mexico.
Santillo DJ, Brown PW, Leslie DM (1989a) Response of songbirds to glyphosate-induced habitat changes on clearcuts. J Wildl Manag 53:64–71.
Santillo DJ, Leslie DM, Brown PW (1989b) Responses of small mammals and habitat to glyphosate application on clearcuts. J Wildl Manag 53:164–172.
Sarmiento L, Llambi LD, Escalona A, Marquez N (2003) Vegetation patterns, regeneration rates and divergence in an old-field succession of the high tropical Andes. Plant Ecol 166:63–74.
Schaffer JD, Sebetich MJ (2004) Effects of aquatic herbicides on primary productivity of phytoplankton in the laboratory. Bull Environ Contam Toxicol 72:1032–1037.
Scribner EA, Battaglin WA, Dietze JE, Thurman EM (2003) Reconnaissance Data for Glyphosate, Other Selected Herbicides, Their Degradation Products, and Antibiotics in 51 Streams in Nine Midwestern States, 2002. 03–217. U.S. Geological Survey, Lawrence, KS.
Skark C, Zullei-Seibert N, Schöttler U, Schlett C (1998) The occurrence of glyphosate in surface water. Int J Environ Anal Chem 70:93–104.
Skark C, Zullei-Seibert N, Willme U, Gatzemann U, Schlett C (2004) Contribution of non-agricultural pesticides to pesticide load in surface water. Pestic Manag Sci 60:525–530.
Smith EA, Oehme FW (1992) The biological activity of glyphosate to plants and animals: a literature review. Vet Hum Toxicol 34:531–543.
Solomon KR, Takacs P (2002) Probabilistic risk assessment using species sensitivity distributions. In: Posthuma L, Traas T, Suter GW (eds) Species Sensitivity Distributions in Risk Assessment. CRC Press, Boca Raton, FL, pp 285–313.
Solomon KR, Thompson DG (2003) Ecological risk assessment for aquatic organisms from over-water uses of glyphosate. J Toxicol Environ Health B 6:211–246.
Solomon KR, Anadón A, Cerdeira A, Marshall J, Sanin L-H (2005a) Environmental and human health assessment of the aerial spray program for coca and poppy control in Colombia. Inter-American Drug Abuse Control Commission (CICAD) section of the Organization of American States (OAS), Washington, DC. http://www.cicad.oas.org/en/glifosateFinalReport.pdf.
Solomon KR, Houghton D, Harris SA (2005b) Nonagricultural and residential exposures to pesticides. Scand J Work Environ Health 31(suppl 1):74–81.

Springborn (2003a) An acute dermal toxicity study in rats with Spray-Charlie. 3596.17. Springborn Laboratories, Spencerville, OH.
Springborn (2003b) An acute nose-only inhalation toxicity study in rats with Spray-Charlie. 3596.18. Springborn Laboratories, Spencerville, OH.
Springborn (2003c) An acute oral toxicity study in rats with Spray-Charlie. 3596.16. Springborn Laboratories, Spencerville, OH.
Springborn (2003d) A dermal sensitization study in guinea-pigs with Spray-Charlie. Modified Buheler design. 3596.21. Springborn Laboratories, Spencerville, OH.
Springborn (2003e) A primary eye irritation study in rabbits with Spray-Charlie. 3596.19. Springborn Laboratories, Spencerville, OH.
Springborn (2003f) An primary skin irritant study in rabbits with Spray-Charlie. 3596.20. Springborn Laboratories, Spencerville, OH.
Springborn (2003g) Purity analysis for glyphosate of Spray-Charlie (active ingredient). 3596.18. Springborn Laboratories, Spencerville, OH.
Stantec (2005a) Mixture of Glyphosate and Cosmo-Flux® 411F: Acute Contact Toxicity to Honey Bee (*Apis mellifera*). Stantec, Guelph, ON, Canada.
Stantec (2005b) Mixture of Glyphosate and Cosmo-Flux® 411F: Acute Toxicity to *Daphnia magna*. Stantec, Guelph, ON, Canada.
Stantec (2005c) Mixture of Glyphosate and Cosmo-Flux® 411F: Acute Toxicity to Fathead Minnow. OECD 203. Stantec, Guelph, ON, Canada.
Stantec (2005d) Mixture of Glyphosate and Cosmo-Flux® 411F: Acute Toxicity to Rainbow Trout. OECD 203. Stantec, Guelph, ON, Canada.
Stantec (2005e) Mixture of Glyphosate and Cosmo-Flux® 411F: Growth Inhibition Test With the Freshwater Green Algae *Selanastrum capricornutum*. Stantec, Guelph, ON, Canada.
Stiltanen H, Rosenberg C, Raatikainen M, Raatikainen T (1981) Triclopyr, glyphosate and phenoxyherbicide residues in cowberries, bilberries and lichen. Bull Environ Contam Toxicol 27:731–737.
Strange-Hansen R, Holm PE, Jacobsen OS, Jacobsen CS (2004) Sorption, mineralization and mobility of *N*-(phosphonomethyl)glycine (glyphosate) in five different types of gravel. Pestic Manag Sci 60:570–578.
Sullivan TP (1994) Influence of herbicide-induced habitat alteration on vegetation and snowshoe hare populations in sub-boreal spruce forest. J Appl Ecol 31:717–730.
Sullivan TP, Sullivan DS (2003) Vegetation management and ecosystem disturbance: impact of glyphosate herbicide on plant and animal diversity in terrestrial systems. Environ Rev 11:37–59.
Sullivan TP, Lautenschlager RA, Wagner RG (1996) Influence of glyphosate on vegetation dynamics in different successional stages of sub-boreal spruce forest. Weed Technol 10:439–446.
Sullivan TP, Nowotny C, Lautenschlager RA, Wagner RG (1998) Silvicultural use of herbicide in sub-boreal spruce forest: implications for small mammal population dynamics. J Wildl Manag 62:1196–1206.
Tadros TF (1994) Surfactants in Agrochemicals. Dekker, New York.
Tate TM, Jackson RN, Christian FA (2000) Effects of glyphosate and dalpon on total free amino acids profiles of *Pseudosuccinae columella* snails. Bull Environ Contam Toxicol 64:258–262.
Tenuta M, Beuchamp EG (1995) Denitrification following herbicide application to a grass sward. Can J Soil Sci 76:15–22.

Terech-Majewska E, Siwicki AK, Szweda W (2004) Modulative influence of lysozyme dimer on defence mechanisms in the carp (*Cyprinus carpio*) and European sheatfish (*Silurus glanis*) after suppression induced by herbicide Roundup. Pol J Vet Sci 7:123–128.

Thomas WE, Pline-Srnic WA, Thomas JF, Edmisten KL, Wells R, Wilcut JW (2004) Glyphosate negatively affects pollen viability but not pollination and seed set in glyphosate-resistant corn. Weed Sci 52:725–734.

Thompson DG, Pitt DG, Staznik B, Payne NJ, Jaipersaid D, Lautenschlager RA, Bell FW (1997) On-target deposit and vertical distribution of aerially released herbicides. For Chronol 73:47–59.

Thompson DG, Wojtaszek BF, Staznik B, Chartrand DT, Stephenson GR (2004) Chemical and biomonitoring to assess potential acute effects of Vision herbicide on native amphibian larvae in forest wetlands. Environ Toxicol Chem 23: 843–849.

Tooby TE (1985) Fate and biological consequences of glyphosate in the aquatic environment. In: Gossbard E, Atkinson D (eds) The Herbicide Glyphosate. Butterworth, London, pp 206–217.

Tsui MTK, Chu LM (2003) Aquatic toxicity of glyphosate-based formulations: comparison between different organisms and the effects of environmental factors. Chemosphere 52:1189–1197.

Tsui MTK, Chu LM (2004) Comparative toxicity of glyphosate-based herbicides: Aqueous and sediment porewater exposures. Arch Environ Contam Toxicol 46:316–323.

United States Department of State (2002) Aerial Eradication of Illicit Coca in Colombia Report on Issues Related to the Aerial Eradication of Illicit Coca in Colombia. United States Department of State. Accessed 2005 April. http://www.state.gov/g/inl/rls/rpt/aeicc/c7470.htm.

UNODC (2003) Global Illicit Drug Trends. E.03.XI.5. United Nations Office on Drugs and Crime, New York. http://www.unodc.org/unodc/global_illicit_drug_trends.html.

Urban DJ, Cook NJ (1986) Standard Evaluation Procedure for Ecological Risk Assessment. EPA/540/09-86/167. Hazard Evaluation Division, Office of Pesticide Programs, U.S. Environmental Protection Agency, Washington, DC.

USEPA (1992) Framework for ecological risk assessment. EPA/630/R-92/001. U.S. Environmental Protection Agency, Washington, DC.

USEPA (1993a) R.E.D. Facts Glyphosate. EPA 738-R-93-014. U.S. Environmental Protection Agency, Washington, DC.

USEPA (1993b) Reregistration Eligibility Decision (RED). EPA 738-R-93-014. U.S. Environmental Protection Agency, Washington, DC.

USEPA (1996a) Ecological effects test guidelines. OPPTS 850.3020. Honey bee acute contact toxicity. U.S. Environmental Protection Agency, Washington, DC.

USEPA (1996b) Ecological Effects Test Guidelines. OPPTS 850.5400. Algal Toxicity, Tiers I and II. U.S. Environmental Protection Agency, Washington, DC.

USEPA (1997) Glyphosate: pesticide tolerances for emergency exemptions. Fed Reg 62:42921–42928.

USEPA (1998) Guidelines for Ecological Risk Assessment. U.S. Environmental Protection Agency, Washington, DC.

USEPA (1999) Glyphosate; Pesticide Tolerance. Fed Reg 64:18360–18367.

USEPA (2001) Environmental Effects Database (EEDB). ECOTOX Database System. U.S. Environmental Protection Agency, Office of Pesticide Programs, Environmental Fate and Effects Division, Washington, DC. Accessed 2005 July 28. http://www.epa.gov/ecotox/.

USNPD (2006) US National Pesticide Use Database. CropLife America. Accessed 2006 March 20. http://www.croplifefoundation.org/cpri_pestuse_2002.asp.

Veiga F, Zapata JM, Fernandez Marcos ML, Alvarez E (2001) Dynamics of glyphosate and aminomethylphosphonic acid in a forest soil in Galicia, north-west Spain. Sci Total Environ 271:135–144.

Vellend M (2003) Habitat loss inhibits recovery of plant diversity as forests regrow. Ecology 84:1158–1164.

Verrell P, Van Buskirk E (2004) As the worm turns: *Eisenia fetida* avoids soil contaminated by a glyphosate-based herbicide. Bull Environ Contam Toxicol 72:219–224.

Walsh LP, McCormick C, Martin C, Stocco DM (2000) Roundup inhibits steroidogenesis by disrupting steroidogenic acute regulatory (StAR) protein expression. Environ Health Perspect 108:769–776.

Wang N, Besser JM, Buckler DR, Honegger JL, Ingersoll CG, Johnson BT, Kurtzweil ML, MacGregor J, McKee MJ (2005) Influence of sediment on the fate and toxicity of a polyethoxylated tallowamine surfactant system (MON 0818) in aquatic microcosms. Chemosphere 59:545–551.

Williams GM, Kroes R, Munro IC (2000) Safety evaluation and risk assessment of the herbicide Roundup® and its active ingredient, glyphosate, for humans. Reg Toxicol Pharmacol 31:117–165.

Williams MR, Abbott I, Liddelow GL, Vellios C, Wheeler IB, Mellican AE (2001) Recovery of bird populations after clearfelling of tall open eucalypt forest in Western Australia. J Appl Ecol 38:910–920.

Wittmann F, Junk WJ, Piedade MTF (2004) The varzea forests in Amazonia: flooding and the highly dynamic geomorphology interact with natural forest succession. For Ecol Manag 196:199–212.

Wojtaszek BF, Staznik B, Chartrand DT, Stephenson GR, Thompson DG (2004) Effects of Vision herbicide on mortality, avoidance response, and growth of amphibian larvae in two forest wetlands. Environ Toxicol Chem 23:832–842.

Wolfenbarger LL, Phifer PR (2000) The ecological risks and benefits of genetically engineered plants. Science 290:2088–2093.

World Health Organization International Program on Chemical Safety (1994). Glyphosate. WHO, Geneva.

Xi YL, Feng LK (2004) Effects of thiophanate-methyl and glyphosate on asexual and sexual reproduction in the rotifer *Brachionus calyciflorus* Pallas. Bull Environ Contam Toxicol 73:644–651.

Manuscript received May 19; accepted June 1, 2006.

Index

MIX
Papier aus verantwortungsvollen Quellen
Paper from responsible sources
FSC® C105338

If you have any concerns about our products,
you can contact us on
ProductSafety@springernature.com

In case Publisher is established outside the EU,
the EU authorized representative is:
Springer Nature Customer Service Center GmbH
Europaplatz 3, 69115 Heidelberg, Germany

Printed by Libri Plureos GmbH
in Hamburg, Germany